Coral Reefs of the World

Volume 22

Series Editors

Coral Reefs of the World is a series presenting the status of knowledge of the world's coral reefs authored by leading scientists. The volumes are organized according to political or regional oceanographic boundaries. Emphasis is put on providing authoritative overviews of biology and geology, explaining the origins and peculiarities of coral reefs in each region. The information is so organized that it is up to date and can be used as a general reference and entry-point for further study. The series will cover all recent and many of the fossil coral reefs of the world.

Prospective authors and/or editors should consult the Series Editors B.M. Riegl and R.E. Dodge for more details. Any comments or suggestions for future volumes are welcomed:

Dr. Bernhard M. Riegl/Dr. Richard E. Dodge Nova Southeastern University Dania Beach, FL 33004 USA

e-mail: rieglb@nova.edu and dodge@nova.edu

Mohammad Reza Shokri • Keivan Kabiri

Coral Reefs of Iran

Biodiversity and Distribution in Extreme Environments

Springer

Mohammad Reza Shokri
Faculty of Life Sciences and Biotechnology
Shahid Beheshti University
Tehran, Iran

Keivan Kabiri
Department of Marine Remote Sensing
Iranian National Institute for Oceanography and Atmospheric Science
Tehran, Iran

ISSN 2213-719X ISSN 2213-7203 (electronic)
Coral Reefs of the World
ISBN 978-3-032-00597-7 ISBN 978-3-032-00598-4 (eBook)
https://doi.org/10.1007/978-3-032-00598-4

This Springer imprint is published by the registered company Springer Nature Switzerland AG
The registered company address is: Gewerbestrasse 11, 6330 Cham, Switzerland

To my beloved father and mother, whose memory continues to inspire me every day. Your love, wisdom, and sacrifices shaped the person I am today, and this book is a tribute to your enduring legacy.
And to my wonderful wife, Saloomeh, for your unwavering support and love, and to my two beautiful daughters, Arnika and Avisa, who bring endless joy and light into my life—you are the heart of my world and my greatest treasures.
This book is for you all, with endless gratitude and love.

—Mohammad Reza Shokri

To my dear father, whose memory I hold close to my heart. Your love, strength, and guidance continue to inspire me in every step I take.
And to my wonderful mother, for your enduring love and support, which has been a constant source of encouragement throughout my journey.
This book is a reflection of your profound influence on my life, and I dedicate it to you with deep gratitude and affection.

—Keivan Kabiri

Foreword

Coral reefs are a biodiverse, but environmentally sensitive, ecosystem. Nowhere is this more evident than in the Persian Gulf, where summer sea temperatures exceed 35 °C, salinity approaches 44 PSU, and coastal development proceeds at a rapid pace. Reefs that persist under these extremes act simultaneously as early-warning systems and as natural laboratories for conditions that much of the tropics will face later this century under climate change.

Iranian reefs anchor this natural experiment. The country's shoreline accounts for roughly 40% of the Gulf's coastal length and spans a marked gradient from the comparatively open Sea of Oman, through the restricted Strait of Hormuz, to the hypersaline inner Gulf. Near-shore frameworks and island-bordering reefs occur along much of this coast, challenging received limits for calcification and coral community persistence. Understanding how these assemblages function, and how they respond to cumulative stress, is important for underpinning regional management and for global projections of coral survival under climate change.

Coral Reefs of Iran: Biodiversity and Distribution in Extreme Environments responds to this need with a synthesis that is empirical, analytical, and solution-oriented. Drawing on field surveys, drone photogrammetry, and multispectral satellite imagery, the authors produce the first integrative map of Iranian reef distribution, establishing a quantitative baseline against which future shifts can be measured. A rigorously compiled taxonomic inventory links each recorded species to abundance trajectories and IUCN status, translating biodiversity documentation into conservation prioritization. Equally valuable is the assessment of stressor intensity, such as land-reclamation, petrochemical discharge, over-extraction, and recurrent marine heatwaves, which are expressed in spatial metrics that managers can incorporate directly into zoning and permitting decisions.

Methodologically, the volume demonstrates how traditional diver-based surveys and modern remote-sensing pipelines can be combined to increase spatial resolution without sacrificing ground-truth accuracy. Validation statistics are reported transparently, fostering reproducibility and encouraging adoption across data-limited jurisdictions. The treatment of monitoring therefore doubles as a temPlate for regional data diplomacy, offering standardized protocols capable of harmonizing assessments across national boundaries.

The implications extend beyond Iran. Documented tolerance ranges refine predictive models of species redistribution under warming seas; quantified ecosystem services strengthen the economic rationale for proactive coastal planning; and the explicit linkage between research outputs and policy frameworks illustrates a pathway from observation to stewardship that other arid-subtropical nations can model.

By integrating distributional mapping, taxonomic rigor, threat analysis, and management guidance, this book moves the discussion of Iranian reefs from anecdote to evidence. It will function as a reference for researchers, a decision-support tool for managers, and a catalyst for coordinated regional action. I commend the authors for delivering a work of both scholarly precision and practical relevance.

Professor of Biology
New York University,
Abu Dhabi, UAE

John A. Burt

Preface

The marine ecosystems of Iran, especially the coral reefs, are rich in biodiversity and hold significant ecological value. This book, *Coral Reefs of Iran: Biodiversity and Distribution in Extreme Environments,* is an attempt to compile comprehensive information about the various species of hard corals found in the Persian Gulf and the Gulf of Oman, along with their distribution.

The coral reefs of Iran, though not as extensively studied as those in other parts of the world, hold significant ecological value. They provide habitat for countless marine species, protect coastlines from erosion, and support local fisheries and tourism industries. However, these vital ecosystems face numerous threats, including climate change, pollution, overfishing, and coastal development. Understanding the diversity and distribution of coral species is crucial for the development of effective conservation strategies.

This book is the result of extensive fieldwork, research, and collaboration with marine biologists, ecologists, and local communities. It aims to serve as a valuable resource for researchers, students, conservationists, and anyone interested in marine biology and the conservation of coral reefs. The detailed descriptions, classification, and photographs of coral species provided herein are intended to aid in the identification and study of these fascinating organisms.

We hope that this book will not only contribute to the scientific understanding of Iran's coral reefs but also inspire efforts to protect and preserve these fragile ecosystems for future generations.

Tehran, Iran

Mohammad Reza Shokri
Keivan Kabiri

Competing Interests The authors have no competing interests to declare that are relevant to the content of this manuscript.

Acknowledgment

Iranian National Science Foundation (INSF), which supported the research expenses under Grant No. 98021239.

Contents

List of Figures

List of Tables

List of Plates

About the Authors

Mohammad Reza Shokri is a marine ecologist specializing in coral reef ecology and conservation in the Persian Gulf and Gulf of Oman. With over 25 years of research experience, he has authored numerous scientific publications and led key projects on coral reef health, biodiversity, and environmental impact assessments, focusing on sustainable marine resource management in Iran.

Keivan Kabiri is an expert in coastal and marine remote sensing, with a focus on habitat mapping, mangrove afforestation, and satellitederived bathymetry. He has extensive experience in integrating field measurements, drone data, and satellite imagery for environmental research. He is currently advancing methods for blue carbon assessment and thermal stress prediction in marine ecosystems.

Introduction

1

1.1 Coelenterata

The term "Coelenterata" encompasses the animal phyla Cnidaria, which includes coral animals, true jellies, sea anemones, sea pens, and their relatives, as well as Ctenophora, commonly known as comb jellies. The designation "Coelenterata" is derived from the Ancient Greek words κοῖλος (koîlos), meaning 'hollow', and ἔντερον (énteron), meaning 'intestine', in reference to the hollow body cavity characteristic of these organisms.

1.2 General Features of Coelenterata

The phylum Coelenterata comprises invertebrate organisms characterized by radial symmetry and the presence of tentacles equipped with nematocysts, or stinging cells. These organisms possess a single internal cavity, the digestive cavity, which has one opening serving as the mouth. A distinctive feature of Coelenterata is their biphasic life cycle, which includes two primary stages: the polyp and the medusa. The polyp, typically cylindrical, is anchored to the substrate with its mouth at the free end, often encircled by tentacles (Fig. 1.1). In both soft and hard corals, polyps may exist either as solitary individuals or as part of a colonial structure, where multiple polyps function collectively to form modular colonies. The medusa, commonly referred to as a jellyfish, features a bell-shaped body, which may be bordered by tentacles (Fig. 1.1). While most jellyfish are free-swimming, some species remain attached to the substrate.

The body wall of Cnidarians, in both their growth forms, comprises three tissue layers: an outer epidermis, an inner gastrodermis, and an intervening gelatinous extracellular matrix known as mesoglea, all of which encase a central body cavity called the coelenteron. The epidermis contains sensory and stinging cells, while the gastrodermis is responsible for digestion and reproduction. The mesoglea, functioning as connective tissue, often includes embedded cells and thus links the epidermis and gastrodermis. This mesoglea layer is particularly well-developed in jellyfish, whereas in other Cnidarians, the body wall is primarily composed of the epidermis and gastrodermis layers.

The first primitive nervous systems, which appeared with the first coelenterates approximately 700 million years ago, consist of one or two networks of diffuse neurons and sensory cells. These sensory cells, functioning as unspecialized general receptors, respond to stimuli such as touch, chemicals, or abrupt temperature changes.

The muscular system in these organisms comprises circular, radial, and longitudinal muscle fibers and layers, which are extensions of the bases of ectodermal and endodermal cells. These muscle fibers and layers enable cnidarians to perform a limited range of motions, including contraction, pumping, and bending.

1.3 Classification of Coelenterata

The extant coelenterates are categorized into four distinct classes: Hydrozoa (hydrozoans), Scyphozoa (scyphozoans), Cubozoa (cubozoans), and Anthozoa (anthozoans) (Fig. 1.2).

The class Hydrozoa, comprising approximately 3000 species, includes colonial medusozoans whose life cycles may involve polyps, medusae, or both. Hydrozoans are unique within the phylum Cnidaria for having freshwater species. They are the only cnidarians that possess both polyp-zooids and medusa-zooids. Colonial species within this class include hydroids resembling seaweed, the pelagic Portuguese man-of-war, and reef-building fire and rose corals. Solitary species in this class consist of marine jellies and the freshwater Hydra.

The class Scyphozoa, encompassing around 200 species across five orders, represents the larger and true jellyfishes, which are notably distinct from the often diminutive hydrozoan medusae. The scyphozoan life cycle includes both polyps and medusae, with medusae being the dominant form and the polyp stage either reduced or absent. The polyp

M. R. Shokri, K. Kabiri, *Coral Reefs of Iran*, Coral Reefs of the World 22, https://doi.org/10.1007/978-3-032-00598-4_1

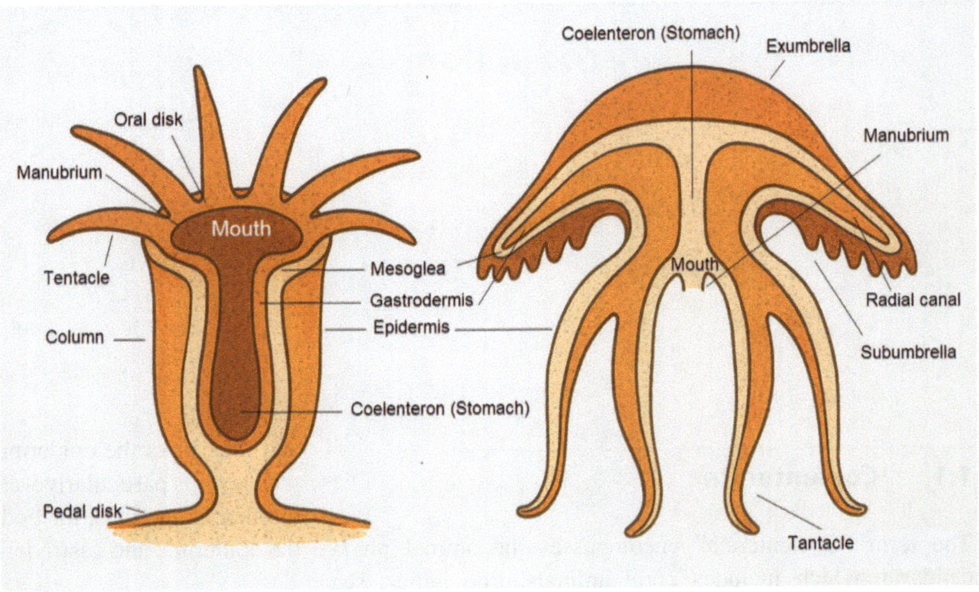

Fig. 1.1 The major growth stages in Coelenterates: Left: Solitary body form (polyp) with radial symmetry. Right: Medusa with radial symmetry

stage, when present, is very small and undergoes strobilation to produce young medusae. Some species entirely lack a polyp stage. While most scyphozoans are marine, a few inhabit freshwater environments. Exclusively marine members of this class include stalked jellies, flag-mouth jellies, and root-mouth jellies.

The class Cubozoa, or box jellies, consists of a small group of approximately 20 marine jellyfish species characterized by a cube-shaped body with four flattened sides, each bearing a tentacle at its corner. The cubozoan life cycle includes both polyp and medusa stages, with the medusa stage being predominant and the polyp stage reduced. The polyp stage develops directly into the medusa. This class includes box jellies and sea wasps.

The class Anthozoa, with about 6000 extant species distributed across nine orders and three subclasses, is entirely marine throughout its life cycle. Anthozoans exhibit either solitary or colonial polyps and lack a medusa stage. The gastric cavity in polyps contains several thin partitions, or septa, which provide additional surface area for digesting large prey. The three subclasses of Anthozoans are Octocorallia or Alcyonaria (e.g., blue coral, soft corals, sea fans, sea pens), Hexacorallia or Zoantharia (e.g., sea anemones, hard corals, zoanthids, mushroom anemones), and Ceriantipatharia (e.g., black and wire corals, tube anemones). The primary distinguishing feature among these subclasses is the variation in the symmetry of polyp structures.

1.4 Subclass Hexacorallia

Hexacorallia, also known as Zoantharia, comprises approximately 4300 species characterized by polyps with typically six-fold symmetry. This subclass of Anthozoa is organized into four extant orders: Actiniaria (sea anemones), Scleractinia (hard corals), Zoanthiniaria (zoanthids), and Corallimorpharia (mushroom anemones). It encompasses all stony corals, most of which are colonial and reef-building, as well as all sea anemones and zoanthids. These organisms are distinguished from another subclass of Anthozoa, Octocorallia, by having six or fewer axes of symmetry in their body structure. The tentacles in Hexacorallia are simple, unbranched, and typically exceed eight in number.

1.5 Order Scleractinia (Hard Corals)

Scleractinia, an order within the subclass Hexacorallia (Zoantharia), encompasses the true or stony corals. These anthozoans can be either solitary or colonial and are typically attached to a firm substrate. They are prevalent in tropical and subtropical waters, playing a significant role in the formation of coral reefs and islands. Some species, however, are free-living and unattached.

The World List of Scleractinia recognizes 1678 valid species of living corals, which may or may not host symbiotic unicellular algae of the genus Symbiodinium. Accordingly, they are classified as zooxanthellate, azooxanthellate, or apozooxanthellate. Corals that harbor these symbiotic algae within their gastrodermal cells are termed zooxanthellate, while those that do not are referred to as azooxanthellate (aposymbiotic). All stony or reef-building (hermatypic) corals are zooxanthellate, whereas those inhabiting deep or cold waters typically lack these symbiotic algae.

Approximately 50% of zooxanthellate coral species are reef-building (hermatypic) and reside in warm, shallow waters. In contrast, the majority of azooxanthellate scleractinians are found in deep, cold waters, with only a few species being stony.

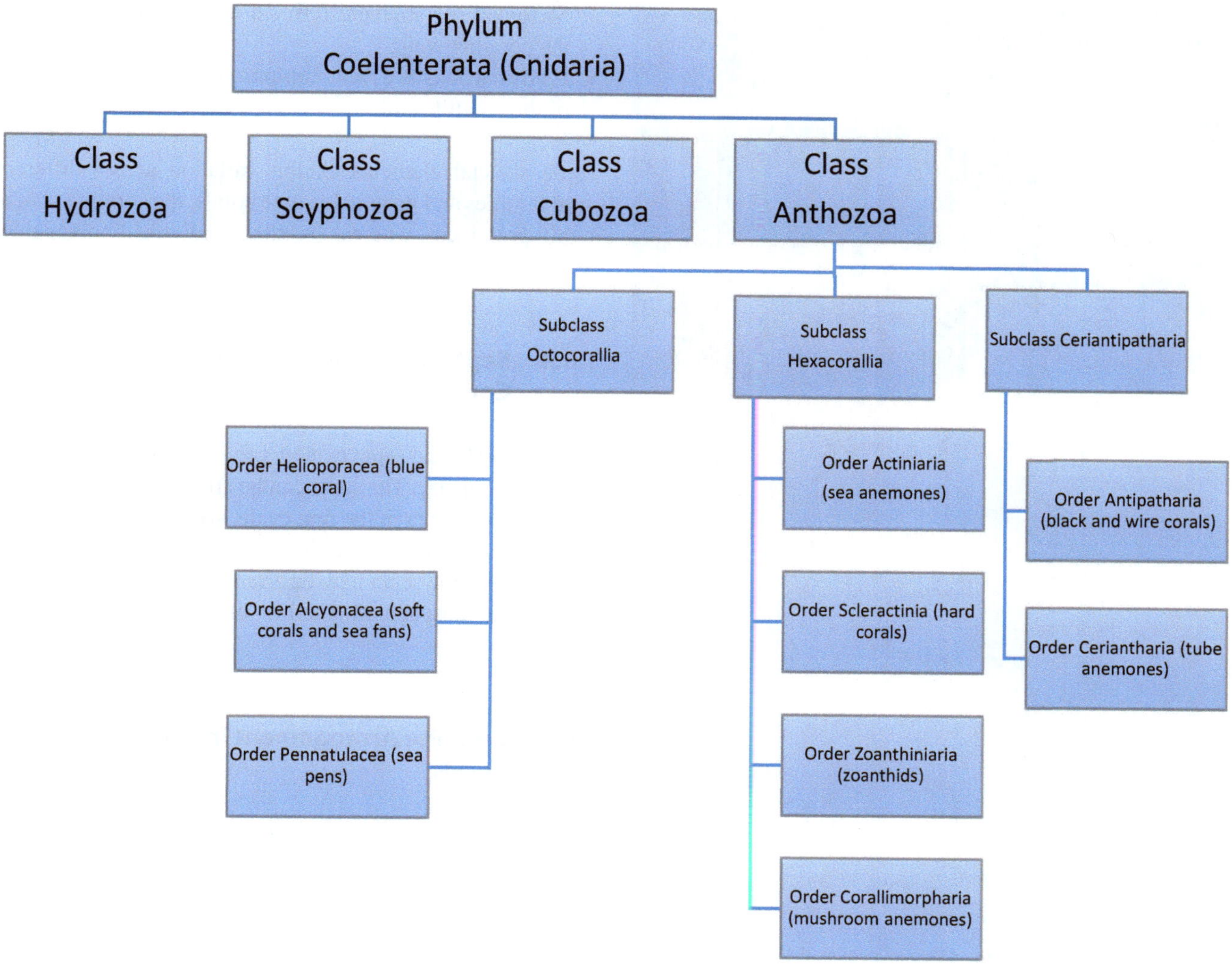

Fig. 1.2 Classification of Coelenterata (Cnidaria)

1.6 Biology of Hard Corals

1.6.1 Coral Anatomy and Structure

Hard or reef-building corals are ancient, primitive animals characterized by a simple body structure with a sac-like cavity and radial symmetry. They possess an opening at the top that functions as both a mouth and an anus. The body wall of hard corals consists of two cell layers: the epidermis and the gastrodermis, separated by a gelatinous connective tissue layer known as the mesoglea. The epidermis contains sensory and stinging cells, while the gastrodermis is responsible for digestion and reproduction. The mesoglea, serving as connective tissue, often contains embedded cells, thereby linking the epidermis and gastrodermis.

The crown structure at the top of each coral, known as a polyp, has a flower-like appearance (Fig. 1.3). The polyp, the primary component of the coral, is a simple sac capped with an oral disc and encircled by a ring of tentacles. In most hard corals, the tentacles remain contracted within the coral skeleton during daylight and expand at night to capture food from the water column. These tentacles are equipped with stinging cells called nematocysts, which are used to capture prey (Fig. 1.3). The corallite, the skeleton of a single polyp, is a tube containing vertical plates called septo-costae that radiate from the center. In colonial corals, adjacent corallites are connected by horizontal plates and other structures collectively known as the coenosteum. Some polyps possess an additional thin skeletal layer around the wall called the epitheca.

1.6.2 Skeletal Elements of Scleractinian Corals

The five primary skeletal elements of scleractinian corals (Fig. 1.4) are as follows:

1. Septo-costae: These are thickened structures within the corallite wall.

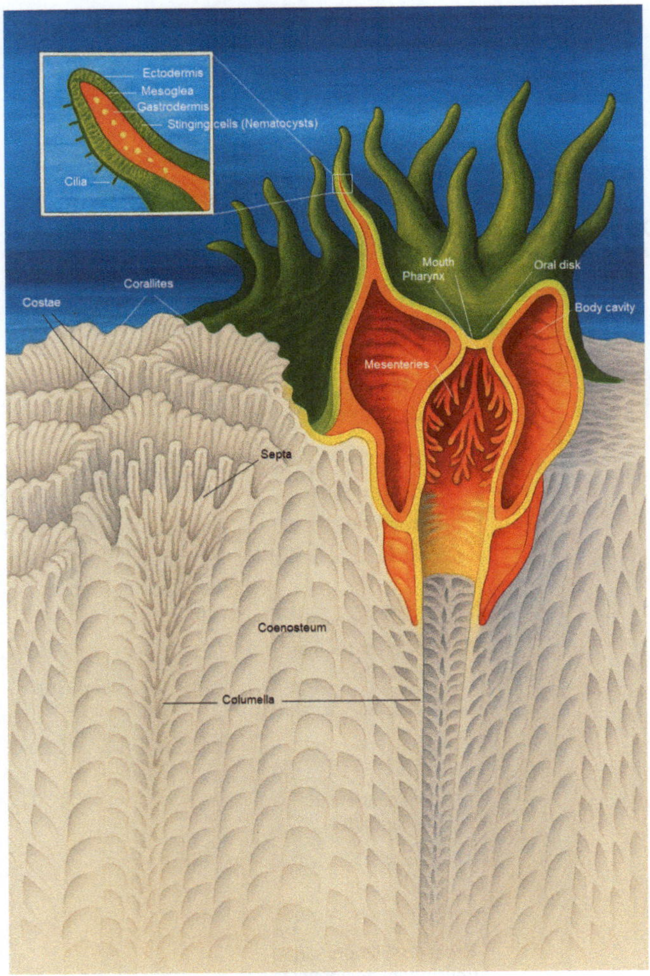

Fig. 1.3 Cut-away view of the polyp in a colonial coral

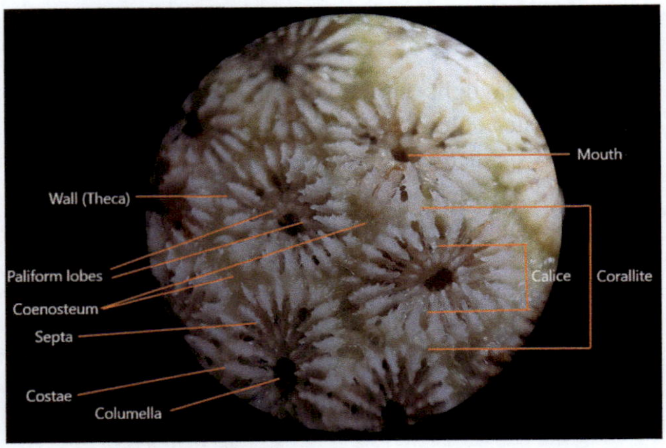

Fig. 1.4 Skeletal elements of corallites in scleractinian coral *Dipsastraea pallida*

2. **Coenosteum:** This refers to the skeletal material between the walls of adjacent corallites, forming a sponge-like structure.
3. Synapticulae: These are horizontal rods that create a network between the septo-costae.
4. Sterome: This element forms a non-porous layer within the corallite wall.
5. Epitheca: This is a thin, non-porous layer on the exterior of the corallite wall.

These skeletal elements exhibit variation across different coral families and genera. In some corals, the corallite wall is highly prominent, whereas in others, individual polyps may be less conspicuous.

1.6.3 Septal Arrangement in Scleractinian Corals

In most corals, the septa exhibit varying lengths and display cyclical symmetry. The septa can be arranged in cycles, typically with six septa in the first cycle, six in the second, 12 in the third, 24 in the fourth, and so on, where present. Alternatively, the septa may be arranged in orders, characterized by an unspecified number of septa of each length (Fig. 1.5).

1.6.4 Corallite Arrangement in Scleractinian Corals

In scleractinian corals, corallites can be arranged in various configurations:

- Plocoid: In plocoid colonies, each corallite possesses its own distinct wall (Fig. 1.6).
- Phaceloid: In phaceloid colonies, corallites possess their own walls, are elongated and tubular, maintain uniform height, and are adjoined at their base (Fig. 1.7).
- Cerioid: In cerioid colonies, corallites share common walls (Fig. 1.8).
- Dendroid: In dendroid colonies, individual corallites are distributed along branches (Fig. 1.9).
- Meandroid: In meandroid colonies, corallite mouths are aligned in valleys with shared walls (Fig. 1.10).
- Thamnasteroid (SeptaCostae): In thamnasteroid colonies, corallites have confluent septa and lack distinct boundaries (Fig. 1.11).
- Flabello-meandroid: In flabello-meandroid colonies, corallites form valleys with multiple mouths and have completely separate walls (Fig. 1.12).

1.6.5 Corallite Positioning in Scleractinian Corals

In scleractinian corals, corallites vary in their positioning on the colony base as follows:

- **Inclined Corallites:** In inclined colonies, corallites grow at an angle, facing the outer margin of the colony base (Fig. 1.13).
- **Exsert Corallites**: In exsert colonies, corallites extend outward or are raised above the colony surface (Fig. 1.14).
- **Immersed Corallites:** In immersed colonies, corallites form depressions inward or are level with or lower than the colony surface (Fig. 1.14).

1.6.6 Growth Forms of Coral Colonies

Coral colonies exhibit diverse growth forms, including: (1) encrusting colonies, which adhere to the substrate as a thin veneer; (2) branching colonies, characterized by the formation of branches; (3) arborescent colonies, resembling tree-like structures; (4) columnar colonies, which develop into columns; (5) laminar colonies, which are plate-like and sometimes form tiers; and (6) massive colonies, which are solid and often hemispherical to spherical, with their underlying solidity derived from the colony's skeleton rather than a veneer on the substrate (Fig. 1.15).

1.6.7 Growth Rates of Coral Colonies

The growth rates of hard corals are species-specific and influenced by environmental conditions. These rates are typically measured as the linear extension of branches, plates, or the radial upward expansion of hemispherical massive corals (Fig. 1.16). Growth rates vary significantly, ranging from approximately 3 mm per year in slow-growing massive corals to over 300 mm per year in fast-growing branching species. Under favorable environmental conditions, such as high light exposure, constant temperature, and moderate wave action, some species can achieve growth rates of up to 4.5 cm per year for slow-growing massive corals and 10 cm per year for fast-growing branching species.

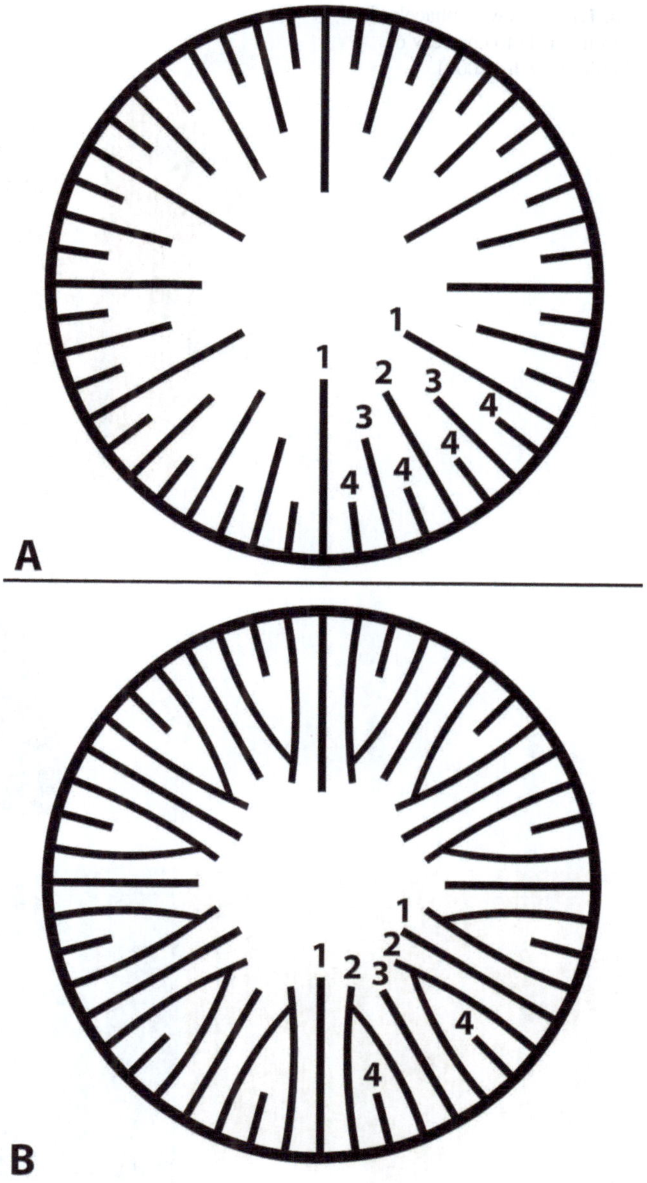

Fig. 1.5 Views of (**a**) the normal cycles of septa and (**b**) pourtalès plan. Numbers represent cycles

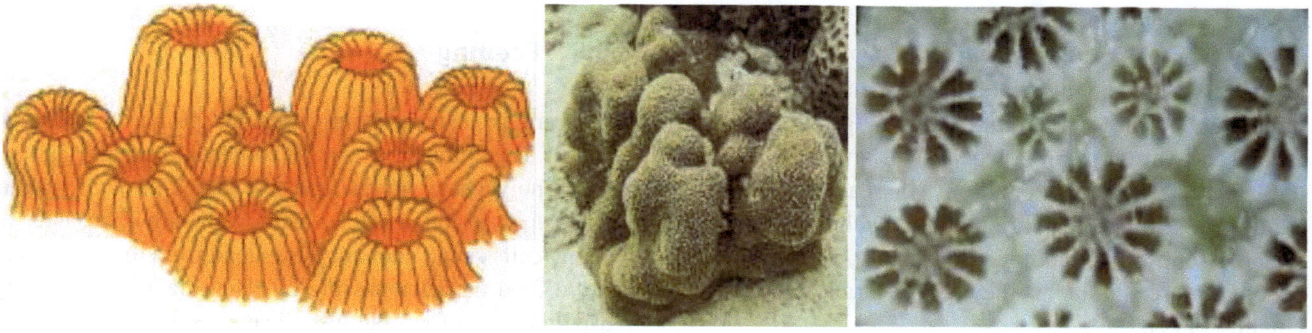

Fig. 1.6 A view of plocoid corallite

Fig. 1.7 A view of phaceloid corallite. (Photo courtesy of Hamid Bargahi, 2009)

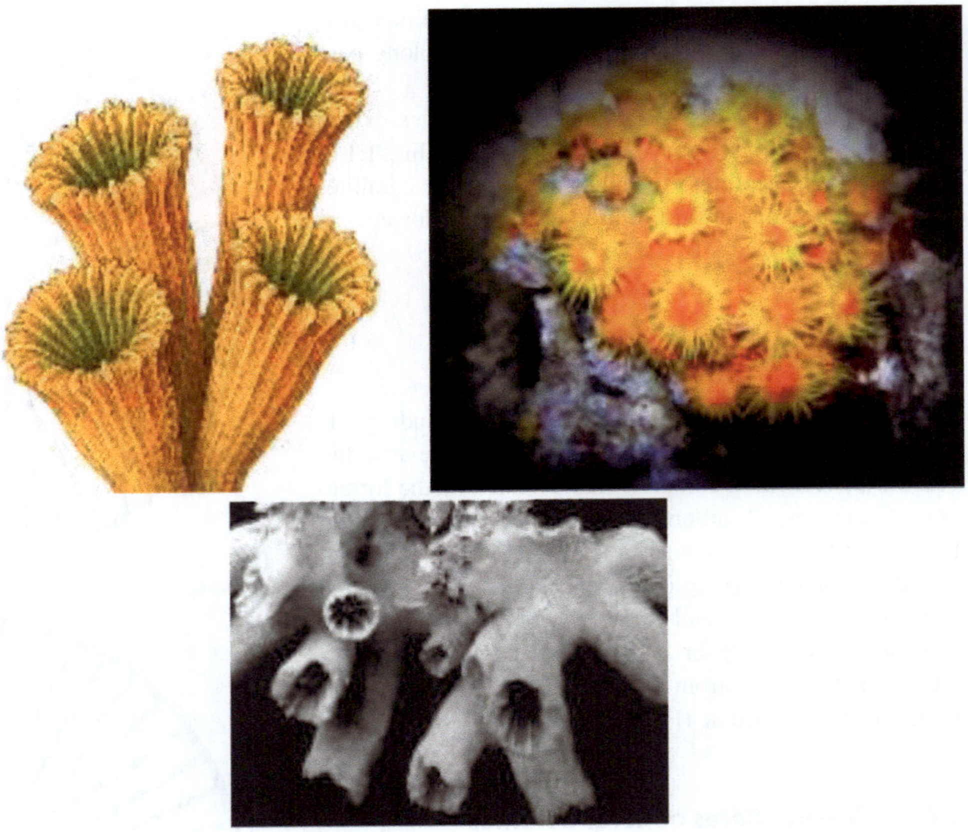

Fig. 1.8 A view of cerioid corallite

However, this rapid growth is not without disadvantages. Mechanical constraints limit the maximum size that branching corals can attain. As they grow larger, the increased load on the relatively small attachment area to the substrate makes the colony increasingly unstable, leading to a higher likelihood of branches breaking off under strong wave action. Conversely, massive corals become more stable as they grow larger (Barnes and Hughes 1999).

1.6.8 Feeding

Corals are exclusively carnivorous and primarily nocturnal feeders. They consume small zooplankton, crab larvae, shrimp, polychaete worms, and fish larvae through various methods. At night, tentacles equipped with thousands of stinging cells called nematocysts extend from their polyps to capture food particles passing by. These tentacles not only grasp and ingest food but also clear debris from the mouth

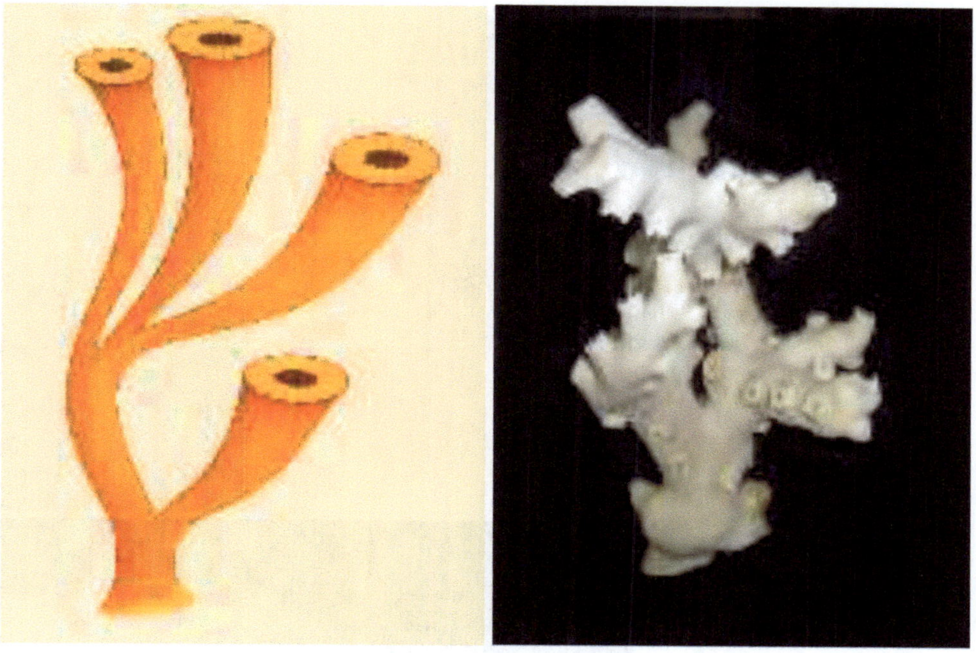

Fig. 1.9 A view of dendroid corallite

and serve as the primary defense mechanism (Barnes 1987; Levinton 2021). The tentacles immobilize prey by piercing it and releasing a lethal toxin, after which the polyp transfers the prey to its mouth.

Corals with shorter, less flexible tentacles secrete slimy mucus to trap dissolved nutrients from water and sediment. The polyp draws the mucus into its mouth using hair-like projections called cilia, which beat cooperatively to create a water current towards the mouth. Long mesenterial threads or filaments can also be extruded through the mouth or body wall to capture food particles and even digest them externally if the prey is too large for the coelenteron or body cavity. Mesenterial filaments are fleshy folds that secrete enzymes to break down and digest food particles in the coelenteron. The digested food particles are then absorbed by the endoderm of the mesenteries. Waste products are expelled in the opposite direction from the mesenteries into the coelenteron and then through the mouth into the water column.

Once the food particles are digested, nutrients can be shared by all polyps throughout the colony. The stomachs of polyps in colonial corals are interconnected by a tissue called coenosarc, which allows the nutrients obtained by individual polyps to be passed to other polyps in the colony via specialized wandering cells called amoebocytes (Fig. 1.17).

1.6.9 Reproduction

Stony corals can reproduce both sexually and asexually (Fig. 1.18). Asexual reproduction occurs through various methods, including budding, fission, accidental and non-accidental fragmentation, partial colony mortality, and polyp bail-out.

Budding reproduction continues throughout the coral's life, with new clonal polyps sprouting from the parent polyp to expand or form new colonies (Fig. 1.19). Budding can occur intra-tentacularly or extra-tentacularly, as well as through fission. In intra-tentacular budding, new polyp buds form from the parent polyp's oral disks, producing polyps of the same size within the ring of tentacles. In extra-tentacular budding, new buds form outside the parent polyp's ring of tentacles, resulting in smaller polyps.

Fission occurs during the early growth stages of some coral colonies, allowing them to divide into two or more colonies (Fig. 1.20). This type of reproduction is common in the Fungiidae family, where solitary mushroom corals can shed or break their skeletons, creating two pieces that then grow into mature corals (Fig. 1.20). Similar reproductive processes occur in Fungiids, which can decalcify part of their skeleton to form acanthocauli, juvenile polyps produced asexually. These juvenile polyps grow on top of their dead parents and eventually become separate polyps.

Fragmentation-based reproduction occurs when a piece of coral is separated from the parent colony, either intentionally or unintentionally (e.g., due to storms, strong wave action, or human disturbance). The detached coral fragment can grow, mature, and form new colonies.

Partial mortality in coral colonies manifests as tissue lesions that are not closed by the surrounding tissue (Fig. 1.21). Corals possess mechanisms that respond immediately to damage, initiating the regeneration of skeleton and tissue in the affected area. Such damage can result from high or prolonged sedimentation, predation, non-lethal diseases,

Fig. 1.10 A view of meandroid corallite

or physical injury. However, recovery from partial mortality imposes a significant energy cost on the coral, potentially disrupting other vital processes that require energy reserves, such as reproduction and growth (Oren et al. 1997, 2001).

Polyp bail-out occurs when an individual polyp abandons its corallite and subsequently settles on the substrate to form a new colony (Fig. 1.22). This phenomenon can be triggered by stressful events such as coral bleaching (Sammarco 1982). The polyp uses two small filaments located at its base to quickly attach to the substrate.

The sexual reproduction cycle in hard corals begins when the coral attains a specific age and size. This process involves the participation of two individuals of opposite sexes, producing gametes (sperm or ova), which can occur within a single colony or across separate colonies. There are four distinct modes of sexual reproduction, varying by species: (1) hermaphroditism with brooding reproduction, (2) hermaphroditism with spawning reproduction, (3) gonochorism with brooding reproduction, and (4) gonochorism with spawning reproduction. In hermaphroditic reproduction, individuals possess both sexes, which can change at a later stage in their life history. Conversely, in the gonochoristic mode, an individual maintains a single sex (male or female) throughout its life. In coral species exhibiting the spawning mode, also known as mass or synchronous spawners, gametes are released into the water column where fertilization occurs, and the fertilized eggs develop into free-swimming larvae called planulae (Fig. 1.23). In brooding species, eggs develop to the planula stage within the gastrovascular cavity of the parent polyp, and once internal development is complete, the planula larvae are released into the water column (Fig. 1.23).

Fig. 1.11 A view of thamnasteroid corallite

Fig. 1.12 A view of flabello-meandroid corallite. (Photo from http://www.reeflifeapps.com)

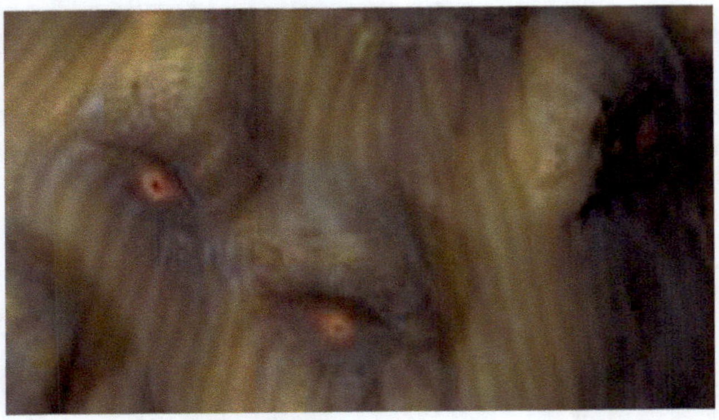

Fig. 1.13 A view of inclined corallites. (Adapted from http://www.reeflifeapps.com)

In tropical regions, coral reproduction can occur year-round (Harrison et al. 1984). In contrast, temperate regions experience synchronized spawning events, often linked to the lunar cycle, where eggs and sperm are released simultaneously (Heyward and Babcock 1986). Certain coral species release eggs and sperm in buoyant bundles, enhancing the concentration of gametes, thereby increasing fertilization success and minimizing self-fertilization risks.

Planulae, the larval stage of corals, typically remain in the water column for several days to weeks (Fig. 1.24). These elliptical larvae have limited swimming capabilities and rely on currents for dispersal. They exhibit some control over

Fig. 1.14 A view of exsert (Left) and immersed (Right) corallites. (Adapted from http://www.reeflifeapps.com)

Fig. 1.15 Growth forms of coral colonies

Fig. 1.16 Growth bands in massive coral *Platygyra daedalea* collected from Hengam Island

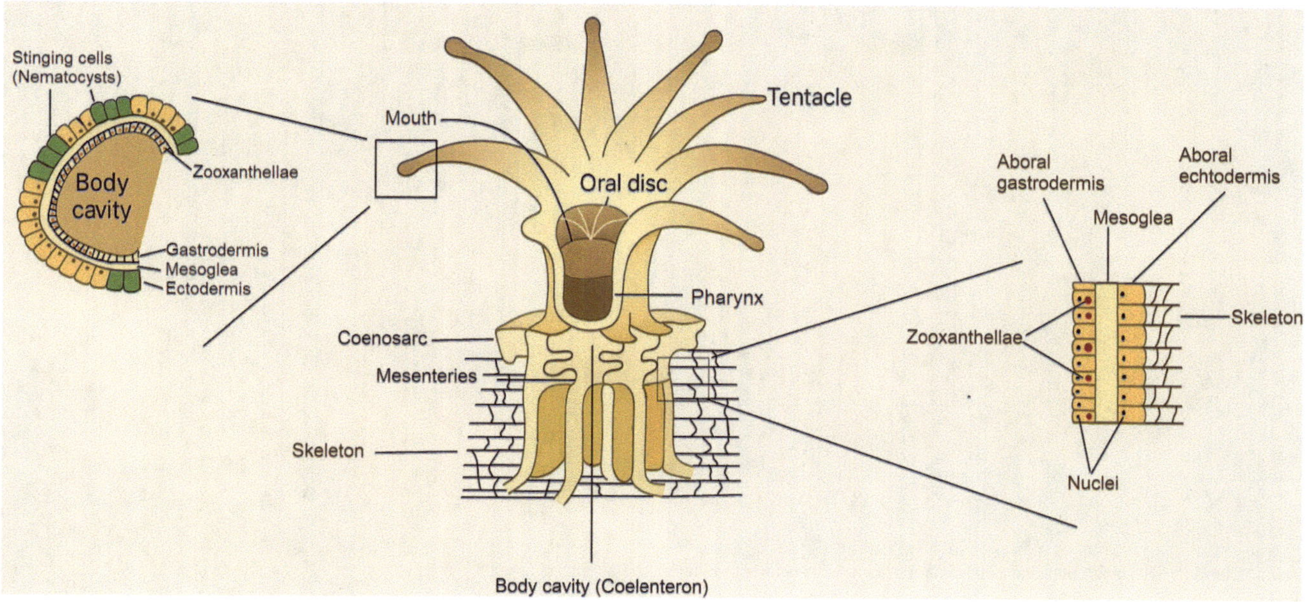

Fig. 1.17 Coral anatomy showing feeding organs

Fig. 1.18 Schematic view of coral life cycle and reproduction

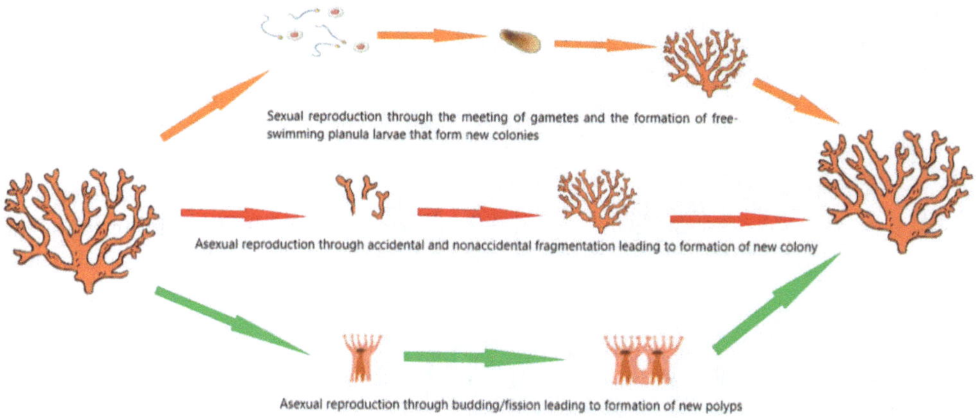

Fig. 1.19 The reproduction by budding: (**a**) Intra-tentacular budding in *Acanthastrea echinata*; and (**b**) Extra-tentacular budding in *Turbinaria reniformis*

Fig. 1.20 Fission in mushroom coral undergoing. (Photo Credit: Khaled bin Sultan Living Oceans Foundation, 2014)

Fig. 1.21 Colony of *Platygyra daedalea* in Kish Island (the Persian Gulf) showing signs of partial colony mortality as a result of bleaching followed by prolonged sedimentation (white portion of the colony)

Fig. 1.22 A view of polyp bail-out. (Photo: Credit: D. Gulko)

their vertical positioning in the water column and demonstrate chemotaxis and phototaxis, which aid in site selection for settlement (Baird and Morse 2004). After the planktonic phase, planulae settle on suitable hard substrates and undergo metamorphosis, secreting a corallite cup from their base to form an adult coral.

1.6.10 Coral Holobiont and Symbiosis

Most reef-building corals engage in symbiotic relationships with a variety of microorganisms, collectively referred to as the coral holobiont. This holobiont includes the coral animal itself and its associated microorganisms, such as bacteria, archaea, fungi, viruses, and protists, including the dinoflagellate algae Symbiodinium, commonly known as zooxanthellae (Fig. 1.25) (Rohwer et al. 2002).

Zooxanthellae are unicellular, yellow-brown dinoflagellate algae approximately one hundredth of a millimeter in diameter, residing in the gastrodermis of corals at densities of up to 1 million cells per square centimeter. Their life cycle alternates between a vegetative cyst and a motile zoospore, with the vegetative stage being predominant in endosymbiosis. These algae perform photosynthesis, similar to other green plants, and release up to 95% of the nutrients produced to the host coral. They also provide pigmentation and a continuous supply of oxygen and other photosynthetic products, such as carbohydrates. In return, zooxanthellae receive a habitat and an abundance of nutrients, including carbon dioxide, nitrogenous, and phosphorus wastes generated from the coral's cellular respiration. This symbiotic relationship is unique, as most corals are voracious feeders on zooplankton, thus having two distinct food sources. However, corals deprived of light for extended periods often die, regardless of the amount of food provided, due to the inability of zooxanthellae to photosynthesize.

Corals acquire their zooxanthellae through different mechanisms, depending on whether they reproduce asexually or sexually. In asexually reproducing corals, zooxanthellae are transmitted via budding or fragmentation, leading to the formation of new corals. In this process, zooxanthellae from the donor tissue of clonal corals are automatically transferred to the new coral. In sexually reproducing corals, zooxanthellae are acquired either through direct (vertical) or

Fig. 1.23 Releasing sperm and egg bundles during mass coral spawning events on the Great Barrier Reef. (Photographs courtesy of the Great Barrier Reef Marine Park Authority)

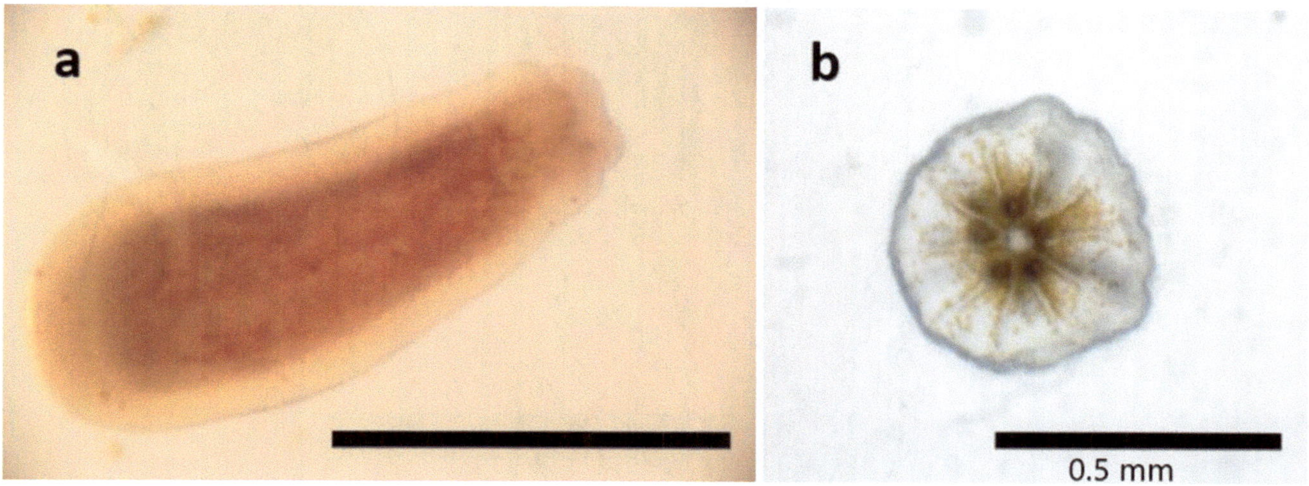

Fig. 1.24 Planula of a brooding species "*Leptastrea purpurea*" (**a**) and primary polyp (**b**). (Photo Credit: Nietzer et al. 2018)

indirect (horizontal) transmission. In direct transmission, corals inherit zooxanthellae from their parent, whereas in indirect transmission, zooxanthellae are obtained from the surrounding environment.

1.7 Ecology of Hard Corals

1.7.1 Distribution

Reef-building corals are predominantly found throughout the tropical and subtropical regions of the Indo-Pacific and Western Atlantic Oceans, typically between latitudes 30°N and 30°S (Fig. 1.26). However, there are some reef-building corals that occur beyond this range, where water currents such as the Gulf Stream bring warm tropical water from the Caribbean north to the islands of Bermuda. The distribution of reef-building corals is primarily constrained to latitudes of 30 degrees north and south due to their symbiotic relationship with unicellular algae (zooxanthellae). This symbiosis necessitates specific environmental conditions to produce the substantial quantities of limestone required for reef construction. Reef-building corals thrive in environments with abundant sunlight, as their symbiotic algae require light for photosynthesis. Consequently, coral growth is restricted to depths where sunlight can penetrate, typically not exceeding 150 feet (approximately 46 meters). Reef-building flourish in oligotrophic (nutrient-poor) waters, with optimal growth

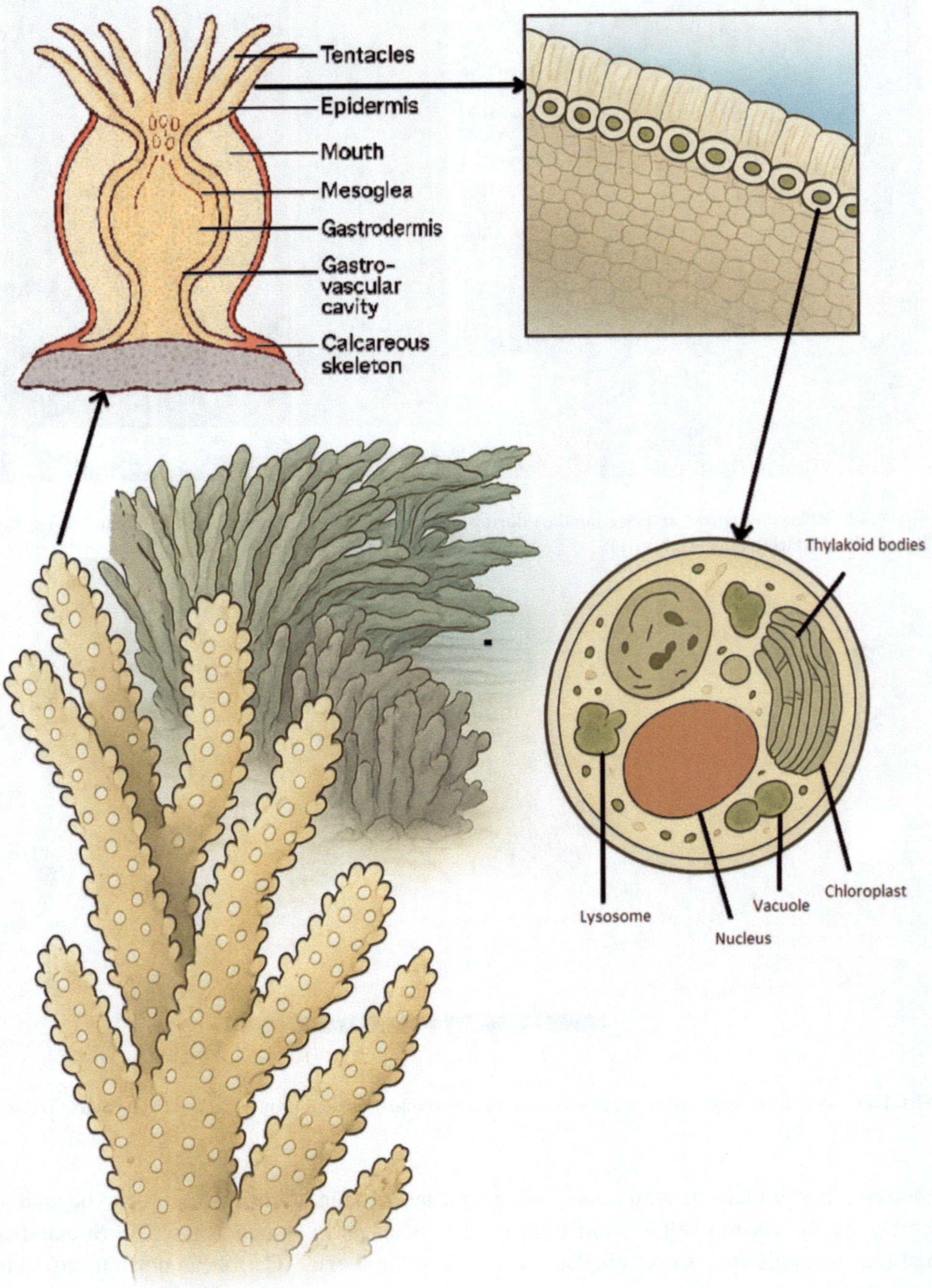

Fig. 1.25 Location of zooxanthellae within the tissue of an individual coral polyp from a coral colony

occurring in salinities of 34 parts per thousand by weight and temperatures ranging from 20 °C (68 °F) to 36 °C (96.8 °F). Additionally, they prefer clear, moving waters that supply essential food, nutrients, and oxygen to the coral polyps.

Coral reefs are estimated to cover approximately 284,300 km^2, representing just under 0.1% of the ocean's surface area (Spalding et al. 2001). About 91% of this area is located in the Indo-Pacific region. Nevertheless, economic assessments of reef value suggest a notably larger estimate of roughly 920,000 km^2 (Costanza et al. 1997). Other estimates, focusing on specific regions, range from half a million to about a million square kilometers, which seems a reasonable approximation of their ecological extent. These discrepancies arise not only from varying measurement methodologies but also from differing definitions of what constitutes a "coral reef" and its associated habitats.

1.7.2 Types of Coral Reefs

Coral reefs are categorized into two main types: continental and oceanic reefs, based on their connection to the mainland.

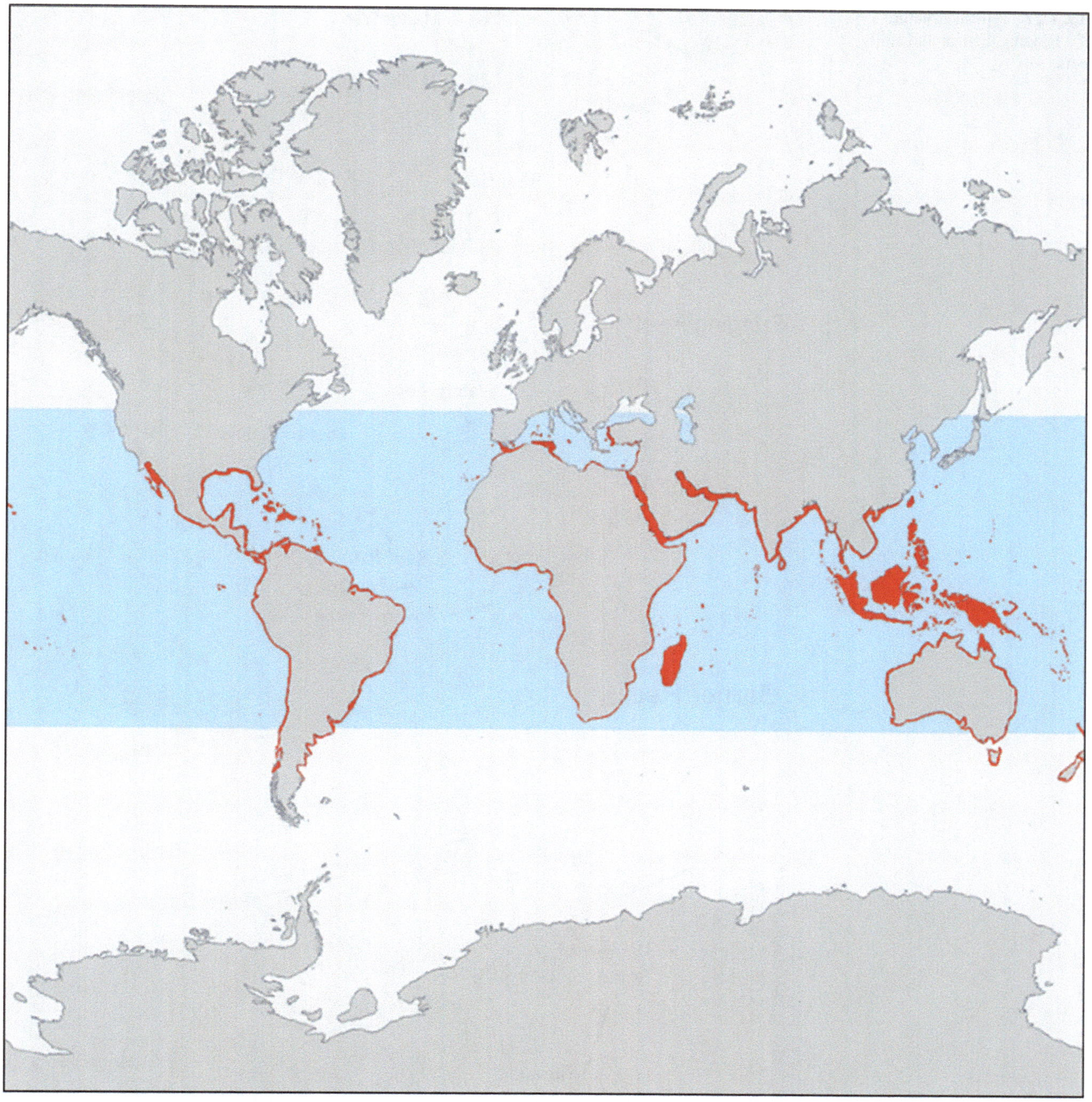

Fig. 1.26 Global distribution of coral reefs. Blue areas indicate areas where zooxanthellate corals occur

Both categories include various reef types such as barrier reefs, fringing reefs, and patch reefs, along with atolls and banks (Figs. 1.27 and 1.28). However, there are distinct differences between oceanic and continental reefs. Oceanic barrier, fringing, and patch reefs are associated with oceanic island reef complexes, whereas continental barrier, fringing, and patch reefs are connected to the mainland.

Within the classification of continental reefs, there are instances where continental islands and their associated reefs are located near the mainland. Atolls and banks are primarily distinguished by the presence or absence of a closing rim and the degree to which their lagoons are enclosed. Unlike atolls and banks, islands do not derive their landmass from carbonate sediments. Fringing reefs form around islands and continental masses, showing variations in their exposure to oceanic swells and their positioning within lagoons and embayments. Barrier reefs, in contrast, are offshore structures separated from the land by lagoons or large sedimentary terraces. Patch reefs, which vary in size, are either intertidal or subtidal structures. Unlike barrier reefs, they are not continuous, and unlike fringing reefs, they are not adjacent to land. These broad reef types can be further subdivided based on their geomorphology, exposure, and depth.

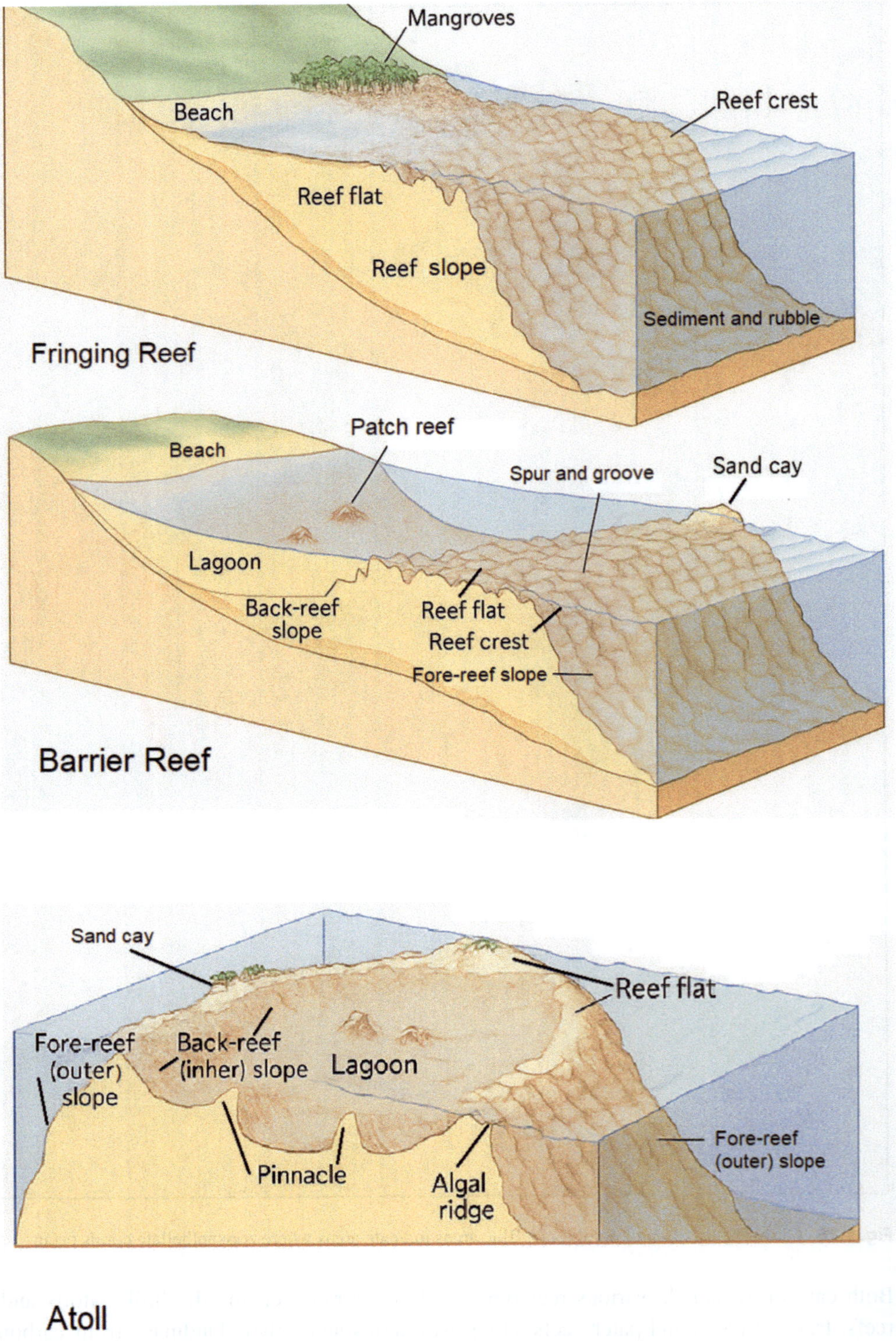

Fig. 1.27 Typical structure of a fringing, barrier and atoll reefs

Fringing reefs are the most prevalent type of coral reef, typically forming along the margins of landmasses and developing primarily in shallow waters, ranging from 40 to 100 meters in depth. These reefs are characterized by an inner reef flat and an outer reef slope, which are separated by a reef crest (Fig. 1.28). The most vigorous coral growth occurs in a narrow band, 20–40 meters wide, at the reef edge or reef front, where continuous surf is present. The outer edge of a fringing reef transitions into areas of soft sediment and rubble. Fringing reefs are commonly found in the Caribbean and around the Hawaiian Islands.

Similar to fringing reefs, barrier reefs run parallel to the shoreline but are often situated some distance from the coast, sometimes extending up to 100 kilometers (60 miles) or more.

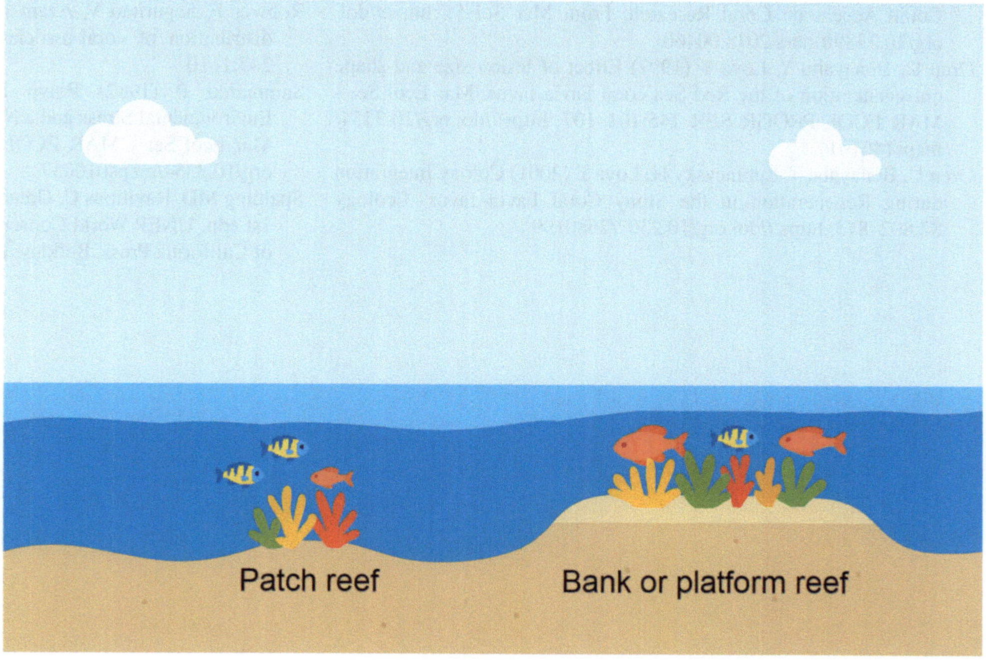

Fig. 1.28 Typical view of a patch and bank or platform reefs

Some barrier reefs are connected to fringing reefs along shelving coastlines, while others develop in offshore locations. These reefs are largely shielded from waves and currents by a deep lagoon with a soft sedimentary bottom that separates them from the shore. Within the lagoon, coral formations such as patch reefs, coral knolls, or pinnacles can be found, which may grow almost to the surface depending on their size and shape. A barrier reef comprises a back reef slope, a flat reef, and a fore reef slope, which corresponds to the fringing reef slope and includes a reef crest. The Great Barrier Reef, located off the eastern coast of Australia, is the world's largest reef system, extending over 2300 kilometers and covering an area of approximately 344,400 square kilometers.

An atoll forms from a fringing reef that develops atop submerged volcanic structures, creating a pyramid-like shape (Fig. 2.2). Unlike fringing and barrier reefs, atolls are typically located in oceanic regions, far from the continental shelf, and rise from depths of thousands of meters. An atoll comprises a ring-shaped reef, often accompanied by islands or sand cays, encircling a central lagoon. Major atolls are predominantly found in the tropical Indo-West Pacific Oceans.

Patch reefs are small, isolated coral formations that are adjacent to each other but physically separated by sand rings (Fig. 1.28). These reefs typically occur in waters 3–10 meters deep and rarely reach the water's surface. They predominantly support large colonies of star and brain corals, with their outer edges often encircled by a ring of sand.

In contrast, bank or platform reefs are found in waters 6–18 meters deep (Fig. 1.28). These reefs are not clearly linked to a coastline and lack the distinct structure of barrier reefs or atolls. Bank reefs are larger than patch reefs and exhibit high species diversity. They often feature spur and groove formations, consisting of low ridges of corals (spurs) separated by sandy channels (grooves). Larger or slightly submerged reef structures of this type are sometimes referred to as shoals.

References

Baird A, Morse A (2004) Induction of metamorphosis in larvae of the brooding corals Acropora palifera and Stylophora pistillata. Mar Freshw Res. https://doi.org/10.1071/MF03121

Barnes RD (1987) Corals and Coral Reefs. In: Invertebrate Zoology, Fifth Edit. Fort Worth, TX: Harcourt Brace JovanovichCollege Publishers, pp 92–96

Barnes RSK, Hughes RN (1999) An introduction to marine ecology, 3rd edn. Oxford Press, Oxford

Costanza R, d'Arge R, de Groot R, Farber S, Grasso M, Hannon B, Limburg K, Naeem S, O'Neill RV, Paruelo J, Raskin RG, Sutton P, van den Belt M (1997) The value of the world's ecosystem services and natural capital. Nature 387:253–260. https://doi.org/10.1038/387253a0

Harrison PL, Babcock RC, Bull GD, Oliver JK, Wallace CC, Willis BL (1984) Mass spawning in tropical reef corals. Science 223:1186–1189. https://doi.org/10.1126/science.223.4641.1186

Heyward AJ, Babcock RC (1986) Self- and cross-fertilization in scleractinian corals. Mar Biol 90:191–195. https://doi.org/10.1007/BF00569127

Levinton J (2021) Marine Biology: Function, Biodiversity, Ecology, 6th editio. Oxford University

Nietzer S, Moeller M, Kitamura M, Schupp P (2018) Coral Larvae Every Day: Leptastrea purpurea, a Brooding Species That

Could Accelerate Coral Research. Front Mar Sci 5:. https://doi.org/10.3389/fmars.2018.00466

Oren U, Benayahu Y, Loya Y (1997) Effect of lesion size and shape on regeneration of the Red Sea coral Favia favus. Mar Ecol Ser - MAR ECOL-PROGR SER 146:101–107. https://doi.org/10.3354/meps146101

Oren U, Benayahu Y, Lubinevsky H, Loya Y (2001) Colony Integration during Regeneration in the Stony Coral Favia favus. Ecology 82:802–813. https://doi.org/10.2307/2680199

Rohwer F, Seguritan V, Azam F, Knowlton N (2002) Diversity and distribution of coral-associated bacteria. Marine Ecol Prog Ser 243:1–10

Sammarco P (1982) Polyp BailOut: An Escape Response to Environmental Stress and a New Means of Reproduction in Corals. Mar Ecol Ser - MAR ECOL-PROGR SER 10:57–65. https://doi.org/10.3354/meps010057

Spalding MD, Ravilious C, Green EP (2001) World atlas of coral reefs, 1st edn. UNEP World Conservation Monitoring Centre. University of California Press, Berkley, USA

Coral Reefs of Iran

2

2.1 Reef Distribution and Development

Iran's coral reefs, encompassing a total area of 700 hectares and constituting 0.25% of the world's reefs (Spalding et al. 2001), are located in two biogeographically distinct regions: the northern Persian Gulf and the northeastern Gulf of Oman (Fig. 2.1, Table 2.1). Despite the differing environmental conditions between the Persian Gulf and the Gulf of Oman, the coral species in these areas belong to the broader Indo-Pacific coral fauna.

The Persian Gulf, with a total area of approximately 240,000 km^2, is a semi-enclosed marginal sea. Situated in the subtropical high-pressure zone, this sea is characterized by low precipitation and high aridity, resulting in evaporation rates exceeding 1500 mm per year (Brewer and Dyrssen 1985). This sea is a shallow sedimentary basin, approximately 100 km in length and between 200 and 330 km in width. The average depth of the Persian Gulf is 38 meters, generally increasing eastward towards the Strait of Hormuz. The deepest areas, located off the Iranian coast, range from 60 meters to about 100 meters at the entrance to the Strait of Hormuz. The entire sea lies within the photic zone. The marine environment of the Persian Gulf is characterized by extreme conditions, with water salinity levels exceeding 45 ppt and highly variable Sea Surface Temperatures (SSTs) ranging from 12 °C in winter to over 36 °C in summer (Reynolds 1993; Coles 2003). As a subtropical epicontinental sea, the Persian Gulf hosts the northernmost coral reefs on the western boundary of the Indo-Pacific. The combination of semi-enclosure, shallow depth, high latitude, and proximity to mountain plateaus and deserts creates the most extreme climate conditions endured by reef-building corals globally (Riegl et al. 2011).

The Gulf of Oman is a deep basin characterized by moderate salinity levels (36–37 ppt) and relatively cooler surface water temperatures, ranging from 20 to 26 °C (Elliott and Savidge 1990; Morrison 1997). This basin is one of the world's five largest upwelling regions, dominated by extensive seasonal upwelling that increases coastal nutrient concentrations during the summer, resulting in a tenfold increase in primary productivity (Schils and Wilson 2006). The reef corals in the Gulf of Oman represent a significantly reduced subset of the Indo-Pacific coral fauna.

The coral reefs in the northern Persian Gulf of Iran are primarily developed as fringing, patch and submereged bank reefs. These reefs are distributed along various parts of the mainland coasts, including Taheri Port, Kangan, and Nayband Bay, all situated in Bushehr Province. Additionally, they are found around Iranian islands and in the Sambarun Bank, which is located at a depth of 13 meters between Kish and Hendorabi islands. In contrast, the patch reefs in Chabahar and Pozm Bays are situated in the northeasternmost part of the Gulf of Oman. Unlike the reefs in the Persian Gulf, which exist near their ecological tolerance limits, the coral assemblages in Chabahar and Pozm Bays benefit from the relatively cooler environmental conditions brought about by monsoonal upwelling.

2.2 Reef Forms

Due to the diverse topographical conditions and relatively harsh physico-chemical environment, Iran's coral reefs exhibit several distinct characteristics, including the absence of typical reef zonation patterns and "spur and groove" structures. Although well-developed reefal framestones are rare, temporary frameworks do occur in the Persian Gulf (Riegl and Purkis 2012). In this sea, reef development primarily takes place in the proximal reaches of the ramp, near the mainland shore, where extensive coast-fringing coral frameworks are found. Consequently, Iran's reefs in the northern Persian Gulf and the Gulf of Oman can be broadly categorized into three forms: (1) Island Fringing Reefs, which are discontinuous and fragmented fringing reefs surrounding

M. R. Shokri, K. Kabiri, *Coral Reefs of Iran*, Coral Reefs of the World 22, https://doi.org/10.1007/978-3-032-00598-4_2

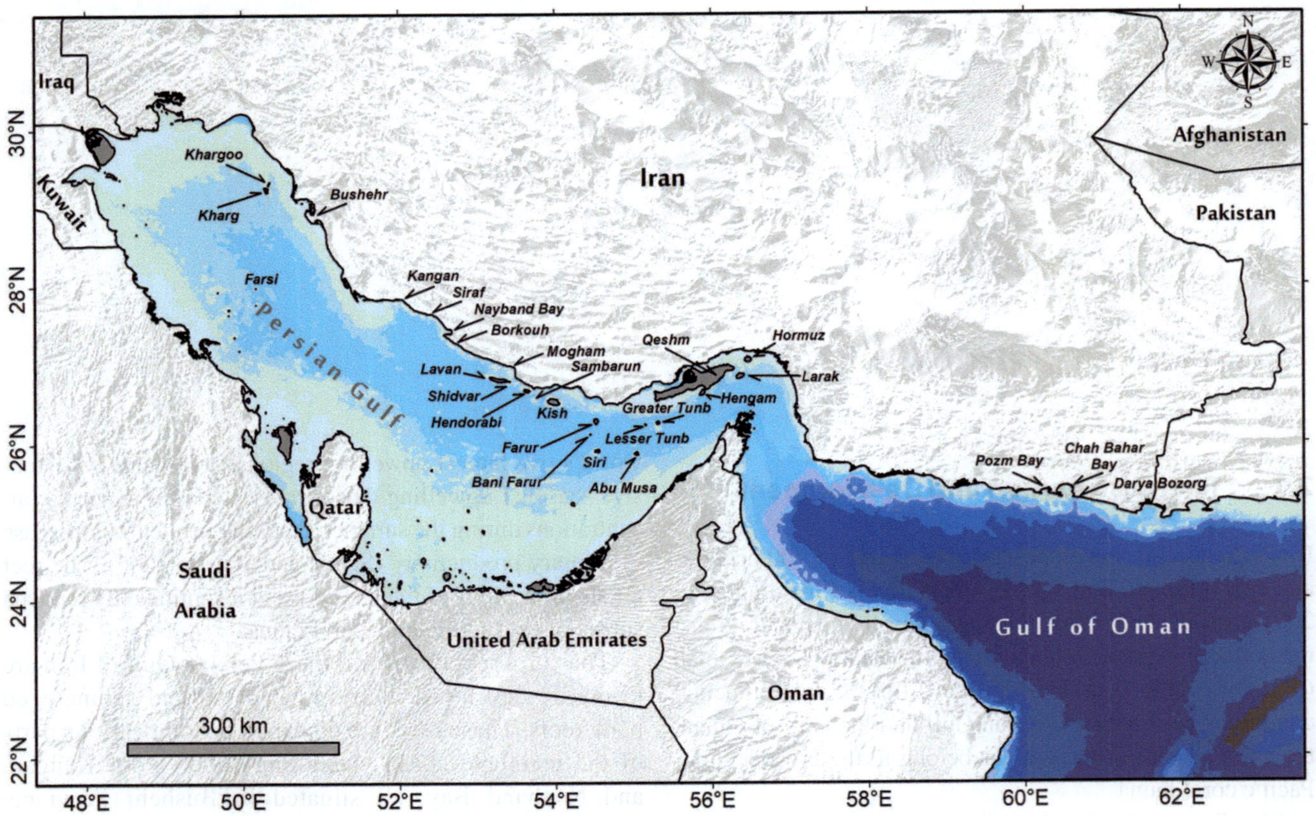

Fig. 2.1 The locations of Iran's coral reefs in the Persian Gulf and the Gulf of Oman

offshore islands, (2) Mainland Fringing Reefs, which are discontinuous and fragmented fringing reefs along mainland shores, and (3) Submerged Bank/Platform Reefs, which are submerged patch reefs developed on offshore banks or platforms.

2.2.1 Island Fringing Reefs

The Iranian islands in the Persian Gulf exhibit variable origins. The island chain off Iran, including Farsi, Kharg, Khargoo, Lavan, Shidvar, Hendorabi, Kish, Farur, Bani Farur, Siri, Abu Musa, Lesser Tunb, Greater Tunb, Hengam, Qeshm, Larak, and Hormuz, is part of the Zagros fold belt (Ross et al. 1986) and is entirely composed of, or at least contains, salt diapirs. These salt diapirs cause continuous uplift, creating terraces of raised reef and shoreline deposits (Pirazzoli et al. 2004).

The islands along the Iranian shores are surrounded by discontinuous and fragmented fringing reefs, which are true coral reefs with extensive biogenic carbonate accretion driven by corals (Plates 2.1 and 2.2). Island fringing reefs typically have gradual slopes with broken-up reef patches and intermittent sand areas. These reefs lack reef crests and typical "spur and groove" structures. The benthic topographical relief, or 'rugosity,' is generally moderate, varying from 1 to 2 meters in height, and sometimes consists of stands of medium-sized massive coral colonies (0.5 to 2 meters in diameter). Live coral cover at these reefs generally ranges from 3 meters to a maximum of 20 meters, with higher coverage at depths of 3 to 8 meters. Due to the counterclockwise pattern of water circulation in the Gulf, the Iranian coastline benefits from better environmental conditions (lower temperature, lower salinity, better aragonite saturation, greater depth), resulting in increased species richness towards the Strait of Hormuz. In general, coral diversity of Iranian island fringing reefs decreases from east to west and from north to south. The islands closer to the mainland generally have lower coral diversity than those offshore. Due to the local climate, intertidal corals are absent in most fringing reefs along the Iranian coastline, likely because high summer air temperatures kill exposed colonies. However, some intertidal corals can be found on Qeshm and Hormuz islands, both located in the Strait of Hormuz, as well as in Chabahar Bay and Tis, both located northeast of the Gulf of Oman.

Table 2.1 Coral reef areas of Iran in the Persian Gulf and Gulf of Oman

Reef name	Reef type	Location	Latitude (°N)	Longitude (°E)	Remarks
Khargoo Island	Island fringing reef	Northwest of the Persian Gulf	29°18′4″	50°19′36″	Uninhabited Island, Wildlife Refuge
Kharg Island	Island fringing reef	Northwest of the Persian Gulf	29°15′0″	50°19′43″	Oil terminal
Farsi Island	Island fringing reef	West of the Persian Gulf	27°59′40″	50°10′5″	Uninhabited Island
Bushehr Mainland	Mainland fringing reef	Northwest of the Persian Gulf	28°57′36″	50°53′24″	Bushehr mainland
Bushehr Mainland (Kangan)	Mainland fringing reef	Northwest of the Persian Gulf	27°50′02″	52°03′46″	–
Bushehr Mainland (Bandar-e Sīraf or Bandar-e Taheri)	Mainland fringing reef	Northwest of the Persian Gulf	27°40′00″	52°20′33″	–
Assaluyeh	Mainland fringing reef	Northwest of the Persian Gulf	27° 28.205″	52° 35.921″	–
Nayband Bay	Mainland fringing reef	Inner north of the Persian Gulf	27°30′47″	52°33′30″	Marine National Park
Borkouh Bay	Mainland fringing reef	Inner north of the Persian Gulf	27°18′21″	52°40′23″	–
Hormozgan Mainland (Mogham Port)	Mainland fringing reef	Inner north of the Persian Gulf	26°59′31″	53°27′29″	–
Lavan Island	Island fringing reef	Inner north of the Persian Gulf	26°48′0″	53°15′0″	Oil terminal
Shidvar Island	Island fringing reef	Inner north of the Persian Gulf	26°47′48″	53°24′23″	Uninhabited Island, Wildlife Refuge
Hendorabi Island	Island fringing reef	Inner north of the Persian Gulf	26°39′36″	53°39′36″	–
Sambarun Bank	Submerged bank/platform reef	Inner north of the Persian Gulf	26°34′0″	53°43′60″	–
Kish Island	Island fringing reef	Inner north of the Persian Gulf	26°32′39″	53°58′34″	Free trade zone
Farur Island	Island fringing reef	East of the Persian Gulf	26°17′15″	54°30′59″	Uninhabited Island
Bani Farur Island	Island fringing reef	East of the Persian Gulf	26°6′55″	54°26′3″	Uninhabited Island
Siri Island	Island fringing reef	East of the Persian Gulf	25°53′38″	54°31′45″	Oil terminal
Abu Musa Island	Island fringing reef	East of the Persian Gulf	25°52′30″	55°1′48″	–
Lesser Tunb Island	Island fringing reef	East of the Persian Gulf	26°14′43″	55°9′4″	–
Grater Tunb Island	Island fringing reef	East of the Persian Gulf	26°14′47″	55°18′32″	–
Hengam Island	Island fringing reef	Strait of Hormuz	26°38′55″	55°51′23″	–
Qeshm Island	Island fringing reef	Strait of Hormuz	26°55′37″	56°15′54″	Free trade zone
Larak Island	Island fringing reef	Strait of Hormuz	26°51′12″	56°21′20″	–
Hormuz Island	Island fringing reef	Strait of Hormuz	27°4′0″	56°27′45″	–
Pozm Bay	Coastal fringing reef	Northeast of the Gulf of Oman	25°19′37″	60°12′26″	–
Chabahar Bay	Coastal fringing reef	Northeast of the Gulf of Oman	25°19′15″	60°37′7″	Free trade and industrial zone, seaport
Darya Bozorg	Submerged bank/platform reef	Northeast of the Gulf of Oman	25°16′40″	60°39′57″	–

Plate 2.1 Shallow reef (4 m) at Abu Musa Island in the Persian Gulf, showcasing a typical island fringing reef

Plate 2.2 Deep reef (14 m) at Kish Island in the Persian Gulf, showcasing a typical island fringing reef

The profile of island fringing reefs generally follows a relatively uniform morphology (Fig. 2.2). Along a vertical section from the shore to the sea, three zones with a relatively gentle slope are visible at island fringing reefs:

1- Sand and Silt Area: At a depth range of 0 to 3 meters, this area is mainly covered by sediments composed of sand and silt with a particle size of 0.63–1.63 mm and coral rubble.
2- Coral Area: At a depth range of 3–20 meters, this area is mainly covered by sand, rocks, and coral colonies. The most turbulent areas in this band are characterized by robust and wave-resistant corals of massive and submassive species (e.g., *Porites*, *Platygyra*, *Dipsastraea*, *Cyphastrea*), high surf areas by wave-resistant morphotypes of branching species (e.g., *Acropora* plates, branching *Acropora*), intermediate depths (e.g., 5–8 meters) by corals with a wide diversity of morphologies, and greater depths (>8 meters) by flattened and plating morphologies (e.g., *Turbinaria*, *Pavona*).
3- Coastal Alluvial Area: Beyond a depth of 18 meters, the seabed is mainly covered by fine sand and silt, mostly with a grain size of 0.63 mm.

2.2.2 Mainland Fringing Reefs

The Persian Gulf coastline of Iran is characterized by steep anticlines formed during the Plio-Pleistocene Zagros Orogeny, displaying dips that transition from 50° inland to 10–20° near the coast (Kassler 1973). Consequently, the Iranian shelf is narrow and bordered by a discontinuous island chain (Ross et al. 1986). Sediments primarily composed of marl dominate the coasts of Iran, influenced by mud and fine sand deposited by rivers in the northwest Persian Gulf. Moving away from the coast, both grain size and carbonate content increase (Seibold et al. 1973). Seasonal turbidity variations near the Shatt al-Arab delta and the mouths of the Hendijan, Hileh, and Mand rivers are believed to

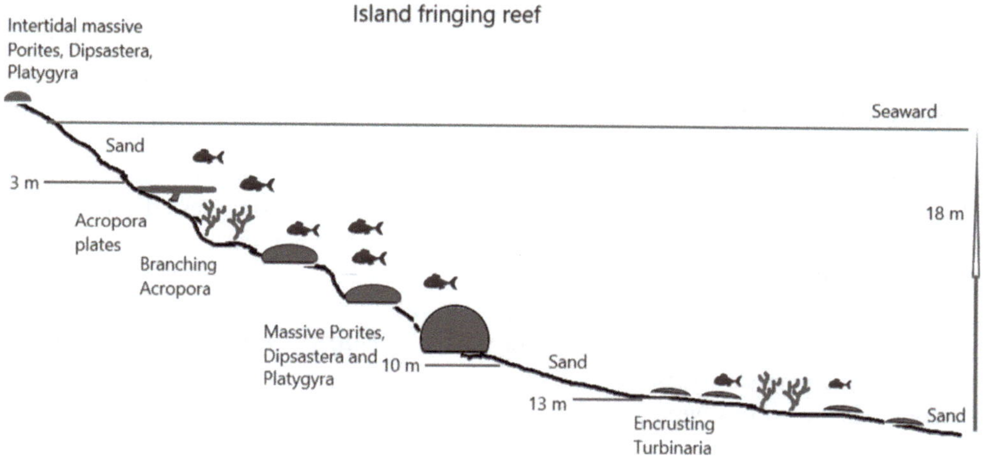

Fig. 2.2 Structural profile and coral assemblages of a typical island fringing reef

inhibit coral reef development in the north Persian Gulf. However, from Kangan to Nayband Bay in Bushehr Province, where the seabed consists of bedrock, there are fragmented fringing reefs along the mainland shorelines.

In contrast, the Gulf of Oman features a deep basin with depths of approximately 200 meters near the Strait of Hormuz, 2000 meters in its central region, and 3000 meters in its eastern part (Reynolds 1993). Its continental shelf is exceptionally narrow, averaging 20 km in width. Unlike the Persian Gulf, the Gulf of Oman is subject to monsoonal influences from the northern Arabian Sea, with cool summer winds from the south and winter winds from the north. High bioerosion rates, driven by the Arabian upwelling system's primary productivity (Rezai et al. 2004), hinder significant reef-building in the Gulf of Oman. Furthermore, much of Iran's coastline in the Gulf of Oman is sedimentary and exposed, rendering it generally unsuitable for coral growth, except in the sheltered bays of Pozm and Chabahar (Rezai et al. 2004).

Key mainland fringing reefs along the Iranian coastline are located in: (1) Bushehr City, Siraf or Taheri Port, Kangan, Nayband Bay (Plate 2.3), and Borkouh Bay, all within Bushehr Province in the northwest Persian Gulf; (2) Mogham Port in Hormozgan Province in the northern Persian Gulf; and (3) Chabahar (Plate 2.4) and Pozm bays in the northeastern Gulf of Oman. The widths of these reefs range from less than 100 meters to 500 kilometers and typically comprise three distinct zones (Fig. 2.3): (1) a sand and silt area at depths of 0–3 meters, mainly consisting of sand and silt sediments and coral rubble; (2) a coral area at depths of 3–8 meters, dominated by robust and wave-resistant corals such as *Porites*, *Platygyra*, *Dipsastraea*, and *Cyphastrea*, and also

Plate 2.3 Shallow reef (5 m) at Nayband Bay in the northern Persian Gulf, showcasing a typical mainland fringing reef

including branching *Acropora* and *Pocillopora*, as well as plating forms like *Turbinaria* and *Pavona*; and (3) a coastal alluvial area beyond 8 meters in depth, mainly covered by fine sand and silt with an average grain size of 0.63 mm.

2.2.3 Submerged Bank/Platform Reefs

The Fig. 2.4 illustrates the structural profile and coral assemblages of a submerged patch reef, extending seaward and reaching depths of approximately 15 meters. The reef begins on sandy substrate with *Acropora* plates, followed by zones of massive coral forms and branching *Acropora*, which provide complex habitats for reef fish. Centrally located are large massive corals like *Porites*, *Dipastrea*, and *Platygyra*, which thrive in moderately deep, stable conditions. Toward the deeper seaward slope, the reef transitions to encrusting *Turbinaria* corals, adapted to lower light levels. The entire reef structure is submerged, unlike fringing reefs, and is surrounded by sand, creating a distinct, isolated coral ecosystem. A small patch of submerged bank/platform reef is located on the offshore Sambarun Bank (Plate 2.5), a submerged platform situated midway between Kish and Hendorabi islands in the northern Persian Gulf. The shallower sections of this reef are typically 13 meters below the sea surface. This submerged reef hosts a moderately diverse coral fauna, predominantly comprising massive and submassive genera (e.g., *Porites*, *Platygyra*, *Dipsastraea*, *Cyphastera*), in addition to wave-resistant morphotypes of branching corals (e.g., *Acropora* plates, branching *Acropora*), and flattened and plating forms (e.g., *Turbinaria*, *Pavona*). Similarly, a small patch of submerged bank/platform reef is also present at Darya Bozorg, located northeast of the Gulf of Oman, at depths ranging from 12 to 15 meters. This submerged reef supports a less diverse coral fauna, primarily dominated by fire corals.

Plate 2.4 Shallow reef (4 m) at Chabahar Bay in the Gulf of Oman, showcasing a typical mainland fringing reef

2.3 Mapping Iran's Coral Growing Areas

This section details the methodological approach for mapping Iran's coral-growing areas in the Persian Gulf and the Gulf of Oman using high-spatial-resolution satellite imagery, including WorldView-2, QuickBird, and GeoEye-1. Previous studies on the use of WorldView-2 and QuickBird imagery for habitat mapping in the Persian Gulf (Kabiri et al. 2013, 2018) have demonstrated the effectiveness of high-spatial-resolution satellite data in distinguishing coral reefs from other benthic substrates. The workflow integrates remote sensing techniques with field-based validation to create detailed thematic maps of benthic habitats. The process encompasses satellite data selection, pre-processing and correction techniques, image classification, field data collection, and accuracy assessment. Additionally, the feasibility of using aerial drone-based photography for shallow-water coral mapping was evaluated.

A combination of optical satellite imagery and in-situ surveys provided a robust dataset for habitat classification. Satellite images were selected based on optimal weather conditions, high water transparency, and minimal atmospheric interference, ensuring the highest possible image

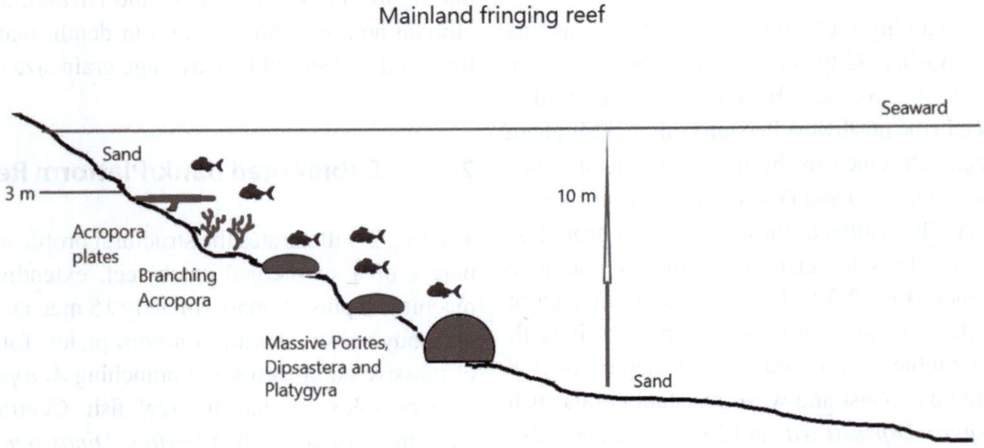

Fig. 2.3 Structural profile and coral assemblages of a typical mainland fringing reef

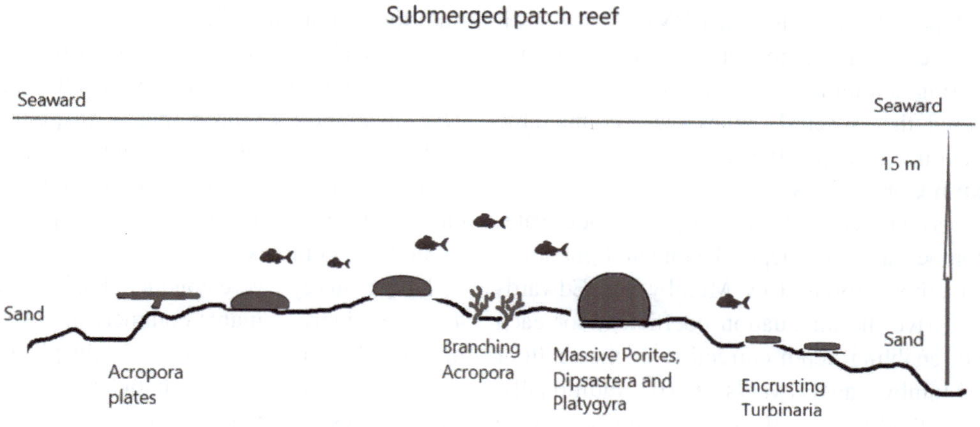

Fig. 2.4 Structural profile and coral assemblages of a typical submerged patch reef

Plate 2.5 Sambarun Bank as a representative example of submerged bank/platform reefs in the northern Persian Gulf

quality for detecting underwater features. Pre-processing techniques were applied to correct geometric distortions, atmospheric effects, and water column interference to enhance the accuracy of habitat classification. Field data, collected through snorkeling-based photography and manta-tow surveys, were used to validate and refine the classification process. The final thematic maps categorized benthic habitats into five classes: (1) sand and rock combinations, (2) sand, (3) gravel and coral rubble, (4) seaweeds, and (5) coral reefs.

2.3.1 Satellite-Based Analysis

2.3.1.1 Selection of Remotely Sensed Data

The selection of satellite imagery was based on several key factors, including water clarity, acquisition timing, and cost considerations. Given that the study areas exhibit high seasonal variability in water turbidity, images acquired during periods of minimal sediment resuspension were prioritized.

2.3.1.2 Pre-processing of Satellite Images

To ensure the highest level of accuracy, multiple correction techniques were applied before classification.

1. **Geometric Corrections:**

   Georeferencing was refined using ground control points (GCPs) collected from GPS field measurements and topographic maps. This process improved spatial accuracy, achieving a final precision of approximately three meters. GeoEye-1 imagery, due to its high inherent accuracy, required minimal additional geometric adjustments. All datasets were projected into the Universal Transverse Mercator (UTM) coordinate system for consistency in spatial analysis.
2. **Radiometric and Atmospheric Corrections:**

   To standardize reflectance values and minimize atmospheric effects, radiometric calibration and atmospheric correction procedures were implemented. First, digital number (DN) values (ranging from 0 to 2047 for 11-bit images) were converted to radiance using ENVI software. The FLAASH (Fast Line-of-sight Atmospheric Analysis of Spectral Hypercubes) module in ENVI 5.6 was then applied to retrieve surface reflectance, correcting for aerosol scattering and molecular absorption effects. This process ensured that spectral values were comparable across different image acquisitions.

3. **Water Column Corrections:**

   One of the primary challenges in mapping underwater habitats is compensating for depth-dependent light attenuation. The method proposed by Mumby and Edwards was used to derive the attenuation coefficient for each spectral band, enabling depth correction of the seafloor reflectance (Mumby and Edwards 2000). Additionally, the Lyzenga method was employed to estimate relative bathymetry in regions with uniform substrates (Lyzenga 1978). These corrections improved the accuracy of benthic classification by accounting for variations in light penetration across different depths.

2.3.1.3 Image Processing and Classification

Image classification was performed using supervised learning approaches, including Maximum Likelihood and Random Forest algorithms. Training data were extracted from field observations, with specific spectral bands selected to optimize class separability. The classification process identified six primary habitat categories: land, sand and rock, sand, rubble, seaweeds, and coral reefs.

Post-Processing and Accuracy Assessment

To enhance classification results, post-processing techniques such as smoothing filters were applied to reduce noise and improve spatial coherence. Thematic maps were validated using accuracy assessment metrics, including confusion matrices and Kappa coefficient calculations. Field-based validation points provided an independent reference for assessing classification performance.

Final Outputs and Applications

The resulting thematic maps provide a comprehensive spatial representation of benthic habitats in Iran's coral-growing regions. These maps serve as valuable resources for ecological studies, conservation planning, and marine spatial management. Additionally, they facilitate the assessment of habitat changes over time, supporting efforts to monitor coral health and resilience in response to environmental stressors.

2.3.2 Drone-Based Analysis

This study also explored the potential of drone-based coral reef mapping in the northern Persian Gulf, specifically targeting Kish, Larak, and Bani Farur islands. The use of unmanned aerial vehicles (UAVs) for coastal habitat mapping offers a cost-effective and high-resolution alternative to traditional field surveys and satellite remote sensing, particularly for shallow-water environments where atmospheric and water column interference limit satellite-based approaches. The integration of drone-based mapping techniques for shallow-water habitat assessment has been explored in recent work (Kabiri et al. 2020), highlighting the potential for enhancing spatial resolution and improving classification accuracy in complex reef environments.

Drone surveys were conducted using the DJI™ Phantom 4 Pro, a widely available commercial UAV equipped with a high-resolution RGB camera. Flight planning and automation were managed through the Pix4Dcapture® software, enabling systematic aerial coverage of each study site. The drone operated at an altitude of 50 meters, ensuring a spatial resolution of approximately 1.3 cm per pixel. AS an instance for Kish Island, the total flight duration per site averaged around 5.5 min, covering an area of approximately 6 hectares (200 × 300 m).

To enhance spatial accuracy, ground control points (GCPs) were established using floating diving buoys, whose coordinates were measured with a handheld GPS. These GCPs were used for georeferencing the drone-acquired images, ensuring high-precision alignment with field data and existing geospatial datasets. The collected aerial imagery was processed using Agisoft™ Metashape to generate high-resolution orthophoto mosaics, which were further corrected and analyzed within ENVI™ 5.3 and AutoCAD® software. The final mosaics were digitized in ESRI™ formats to delineate different coral species and substrate types, including patches of bleached corals.

The key advantages of the UAV-based mapping approach include:

- **Superior spatial resolution:** The ~1.3 cm pixel resolution allows for the identification of fine-scale reef structures and individual coral colonies, surpassing the capabilities of satellite imagery.
- **Cost-effectiveness:** Compared to extensive field-based surveys, drone mapping reduces logistical costs while maintaining high accuracy.
- **Rapid data acquisition:** A single flight can map an extensive area within minutes, facilitating frequent monitoring and change detection.
- **Differentiation of coral conditions:** The method successfully identified and mapped live corals, rubble, seaweeds, and sand, as well as distinguishing bleached corals from healthy ones.

The results from this study align with previous research demonstrating the effectiveness of drone-based approaches for shallow-water coral mapping in the Persian Gulf (Kabiri et al. 2020). While the method provides high spatial accuracy, limitations include its reliance on optimal weather conditions and the lack of multispectral imaging capabilities, which would enhance habitat classification. Future applications may integrate multispectral drone sensors to improve classification accuracy and automate substrate differentiation.

2.4 Iran's Coral Growing Areas in the Persian Gulf

2.4.1 Kharg Island

Kharg Island, also known as Kharg Island, a designated wildlife refuge, is encircled by coral reefs, with the densest coral formations along the northern and eastern shores (Fig. 2.5). The reefs are limited to shallow waters (<8 m) and are relatively rich and diverse, previously home to one of the most abundant *Acropora* communities. Following concurrent bleaching events, the reefs are now dominated by heat-tolerant corals such as *Cyphastrea*, followed by *Platygyra* and *Porites* (Plate 2.6). The southwestern part of the island has reduced coral cover, characterized mainly by rock and rubble. The coral reefs on Kharg Island are threatened by fishing activities and pollution from the oil terminal and desalination plant. There are no specific restrictions on marine life removal beyond the generally applicable national or subnational regulations.

2.4.2 Khargoo Island

Khargoo Island, designated as a wildlife refuge, features dense coral reef coverage along its northeastern and southeastern coasts (Fig. 2.6). In contrast, the western part of the island comprises a mix of sand and rubble. Similar to Kharg Island, the reefs around Khargoo Island are restricted to shallow waters (<8 m) and exhibit considerable richness and diversity. Previously home to one of the most abundant *Acropora* communities, the reefs are now dominated by heat-tolerant corals such as *Cyphastrea*, followed by *Platygyra* and *Porites*, following successive bleaching events.

2.4.3 Farsi Island

Coral reefs on Farsi Island are primarily located along the northern and southeastern coasts, whereas the western shores are mainly characterized by rubble and sand (Fig. 2.7). Due to its offshore location and status as a military restricted zone, access to the reefs on Farsi Island is limited, which has helped maintain their relatively pristine condition. However, successive bleaching events have adversely affected the island's coral reefs, leading to a significant loss of *Acropora* corals (Plate 2.7).

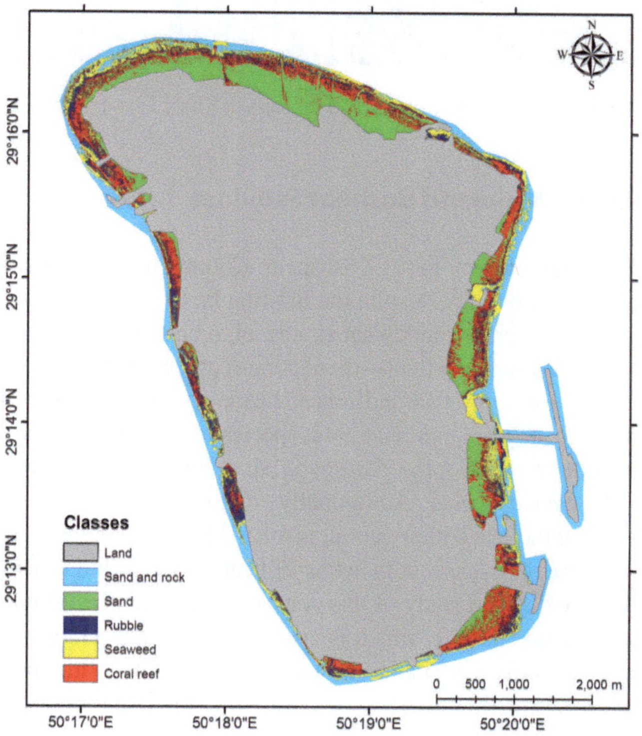

Fig. 2.5 Detailed map of substrate classes around Kharg Island

Plate 2.6 Underwater views of coral reefs in Kharg Island

2.4.4 Mainland Bushehr Province

The southeastern part of Bushehr County hosts scattered patchy coral reefs. Despite the limiting factors for hard coral growth—such as muddy substrates, high nutrient loads from rivers, and reduced light—these coral reefs provide a suitable habitat for many fish and other marine species during their juvenile stages. Dominant coral species in these patchy reefs include *Porites* spp., *Platygyra* sp., *Turbinaria* sp., and *Tubastrea coccinea*. Additionally, *Paracyathus stokesii*, an azooxanthellate solitary coral, is relatively common throughout these reefs, especially in low-light areas and turbid waters. A more detailed study of this coral patch is essential for preserving this habitat, which is crucial for the conservation of marine biodiversity in the coastal waters of Bushehr Province.

2.4.5 Kangan, Assaluyeh, Nayband and Borkouh Bays

The coral reefs in Nayband Bay and the surrounding areas, including Borkouh Bay to the east of Nayband Bay and the patchy reefs extending to Siraf Port (also known as Taheri Port) and Kangan County, are classified as mainland fringing reefs. Coral patches between Kangan and Nayband Bay are primarily observed in two areas: within Nayband Bay itself and between the piers of Assalueyh and Nakhl Taqi, extending to Kangan and Siraf Port (Fig. 2.8). Sporadic studies in these regions indicate that over 90% of the corals between Assalueyh and Nakhl Taqi, as well as between Kangan and Siraf Port, have suffered bleaching and subsequent mortality due to rising water temperatures. These corals now exist as small, scattered colonies at depths of 3–6 meters. Recent decades' activities, particularly construction and land reclamation related to the development of petrochemical complexes in Assalueyh and Kangan, have inflicted significant damage on the coral reefs in these areas. In contrast, the corals in Nayband and Borkouh bays, which extend to greater depths (3–8 meters), are in better condition regarding colony size, species diversity, and health. The overall coral diversity from Borkouh Bay to Kangan is very low. Branching *Acropora* corals, which thrive in low-stress environments, are largely absent in these areas, with only a few newly emerging specimens.

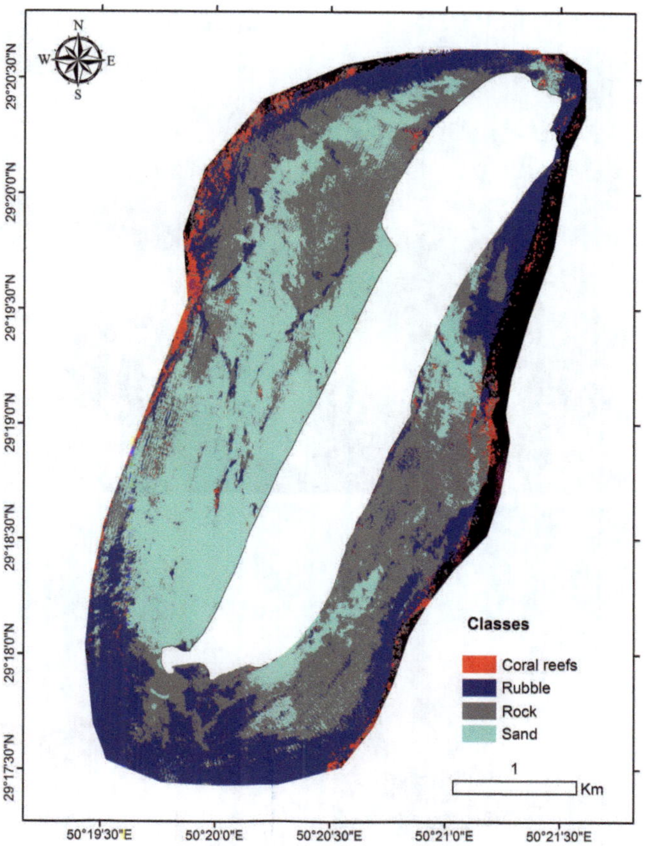

Fig. 2.6 Detailed map of substrate classes around Khargoo Island

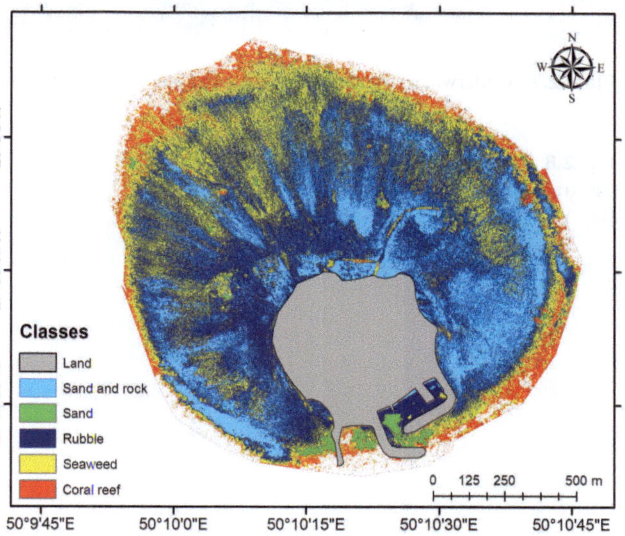

Fig. 2.7 Detailed map of substrate classes around Farsi Island

Plate 2.7 Underwater views of coral reefs in Farsi Island. (Photos courtesy of Hamid Rezai Marnani)

Fig. 2.8 Detailed map of substrate classes in Nayband Bay

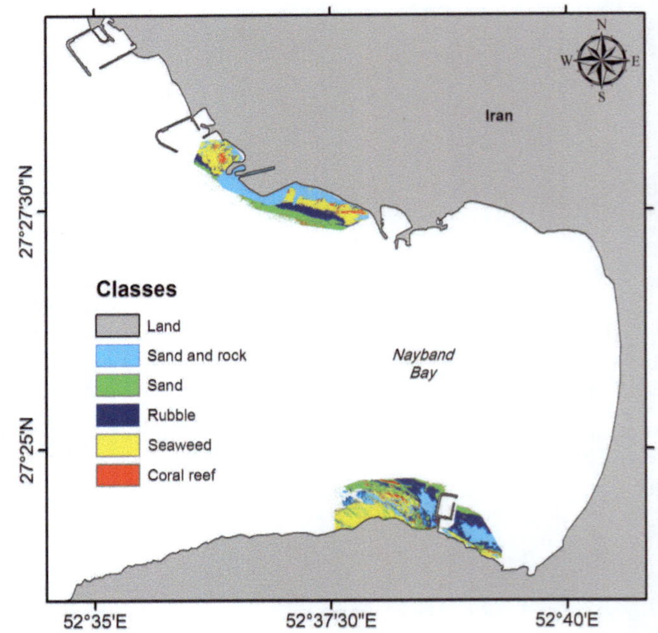

The remaining reef-building corals are limited to genera such as *Porites*, *Cyphastrea*, *Platygyra*, *Favites*, and *Dipsastraea*, all of which exhibit massive and semi-massive growth forms adapted to high-stress environments (Plate 2.8). Evidence suggests that, in the past, coral coverage extended more broadly to the northern parts of the studied region. However, coastal construction and the discharge of various effluents from desalination plants have led to extensive destruction of these corals. Consequently, no live corals remain beyond Nakhl Taqi pier, with most having died years ago.

2.4.6 Lavan Island

The coral reefs surrounding Lavan Island exhibit a distinct distribution pattern, shaped by unique environmental factors of the region. Coral coverage is predominantly concentrated along the northern, southeastern, and southwestern coastlines, where water movement and nutrient levels create favorable growth conditions (Fig. 2.9). The distribution is heterogeneous, with denser coral assemblages occurring in sheltered zones that provide protection from strong tidal forces. The northern coastline of Lavan Island hosts coral reefs in relatively favorable condition, with improved ecological parameters, optimal water transparency, and suitable substrate for coral expansion. Ecologically, the northern sector of the island provides an environment conducive to coral proliferation. In terms of coral species diversity, branching corals of the genus *Acropora* are dominant, followed by massive and sub-massive growth forms, including species of *Porites*, *Platygyra*, *Dipsastraea*, and *Cyphastrea* (Plate 2.9). These reefs contribute substantially to marine biodiversity, offering essential habitats and feeding grounds for a variety of fish and invertebrate species. However, certain portions of the coral reefs have been adversely affected by human activities, such as shipping and coastal development, resulting in degraded coral sections, especially near industrial areas. The observed variability in coral coverage underscores both the resilience

Plate 2.8 Underwater views of coral reefs in Assalueyh, Nayband and Borkouh bays

Fig. 2.9 Detailed map of substrate classes around Lavan Island

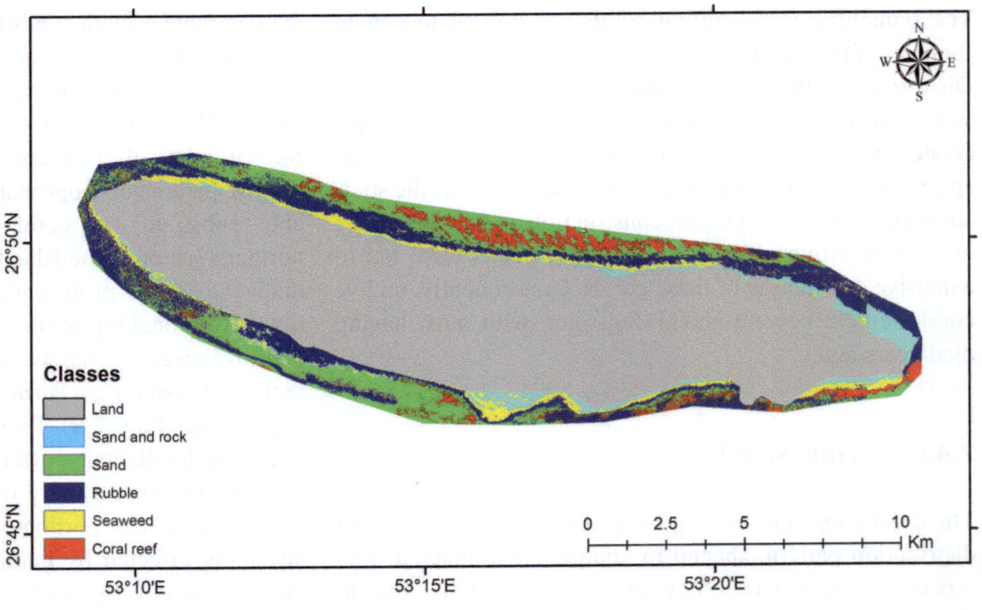

Plate 2.9 Underwater views of coral reefs in Lavan Island. (Photos courtesy of Hamid Reza Bargahi)

and the susceptibility of these ecosystems to natural and anthropogenic pressures.

2.4.7 Shidvar Island

Shidvar Island, also known as Marou, is recognized as a protected coral reef area, designated as a Ramsar Site and Wetland of International Importance, located in the Persian Gulf. This designation highlights its crucial role in maintaining the region's ecological balance and conserving the diverse animal and plant species present. The island's coral reefs cover the western and southern shores, while the eastern and northern shores are dominated by rock and rubble (Fig. 2.10). Branching corals of the genus *Acropora* are dominant, followed by massive and sub-massive growth forms, including species of *Porites*, *Platygyra*, *Dipsastraea*, and *Cyphastrea*. There are no specific restrictions on marine life removal beyond generally applicable national or subnational regulations.

2.4.8 Hendorabi Island

Coral reefs on Hendorabi Island are distributed around the island, with significant concentrations along the eastern and southeastern coasts (Fig. 2.11). The western part of the island predominantly features a mix of rubble and sand, with a reduced coral presence compared to the eastern side. Most corals are found at depths between 1 and 6 meters, with the most commonly observed corals belonging to the genera *Acropora*, *Pavona*, *Dipsastraea*, *Porites*, and *Platygyra* (Plate 2.10). The primary threat to these reefs is coastal construction. In recent years, the construction of an airport and a jetty has led to increased turbidity in the seawater, adversely affecting the coral reefs.

2.4.9 Kish Island

The coral reefs of Kish Island are predominantly located along the northern coastline, with substantial coral patches on the northeastern periphery (Fig. 2.12). These reefs, composed of both live and dead coral structures, are encircled by rock formations and coral rubble. In contrast, the western and southern coastlines exhibit reduced coral cover, predominantly characterized by sand and rocky substrates. The primary threats to these coral reefs include ecotourism and minor coastal developments. Coral reefs around Kish Island occur as discontinuous fringing reefs at depths of 3–18 metersn (Plate 2.11). Due to the proximity of major coral reefs to the beach (approximately 10 meters), SCUBA divers can reach the corals by shore diving. However, amateur divers are strongly recommended to undertake boat diving. The presence of hotels and diving clubs in the eastern part of the island leads to high visitor traffic, causing saturation of the reefs. Most of the recorded damage is attributed to diving-related activities. The most turbulent areas are characterized by robust, wave-resistant corals of massive and submassive corals, such as *Porites*, *Platygyra*, and *Dipsastraea*. High surf areas feature wave-resistant morphotypes of branching genera, like *Acropora*. Intermediate depths (5–8 meters) host corals with a wide diversity of morphologies, while greater depths (>8 meters) are dominated by flattened and plating morphologies, such as *Turbinaria* and *Pavona*. Beyond 18 meters, the seabed is mainly covered by fine sand and silt, with a typical grain size of 0.63 mm.

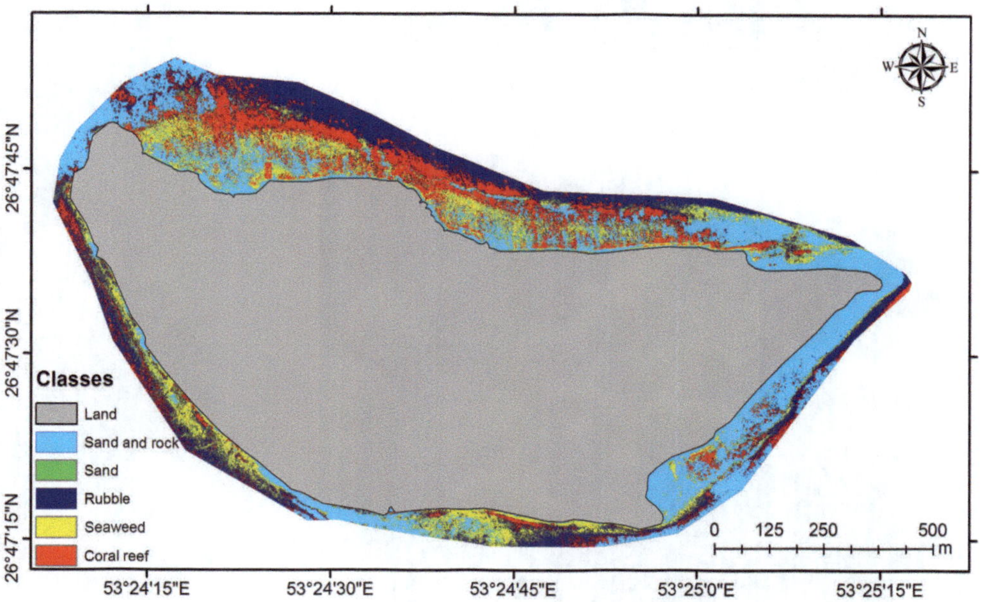

Fig. 2.10 Detailed map of substrate classes around Shidvar Island

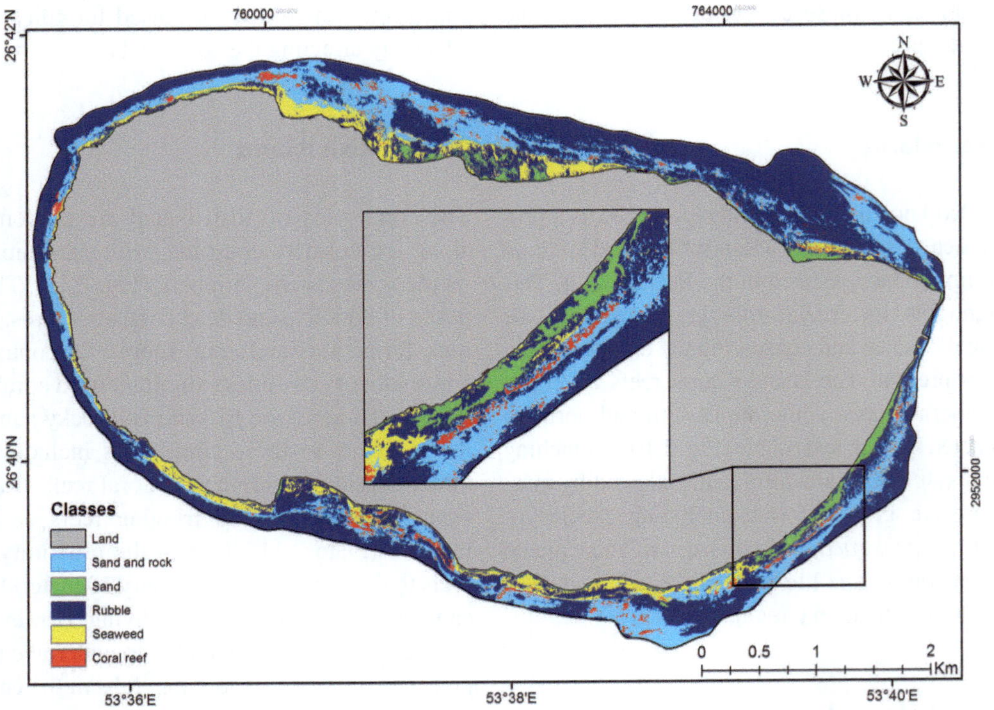

Fig. 2.11 Detailed map of substrate classes around Hendorabi Island

Plate 2.10 Underwater views of coral reefs in Hendorabi Island. (Photos courtesy of Hamid Rezai Marnani)

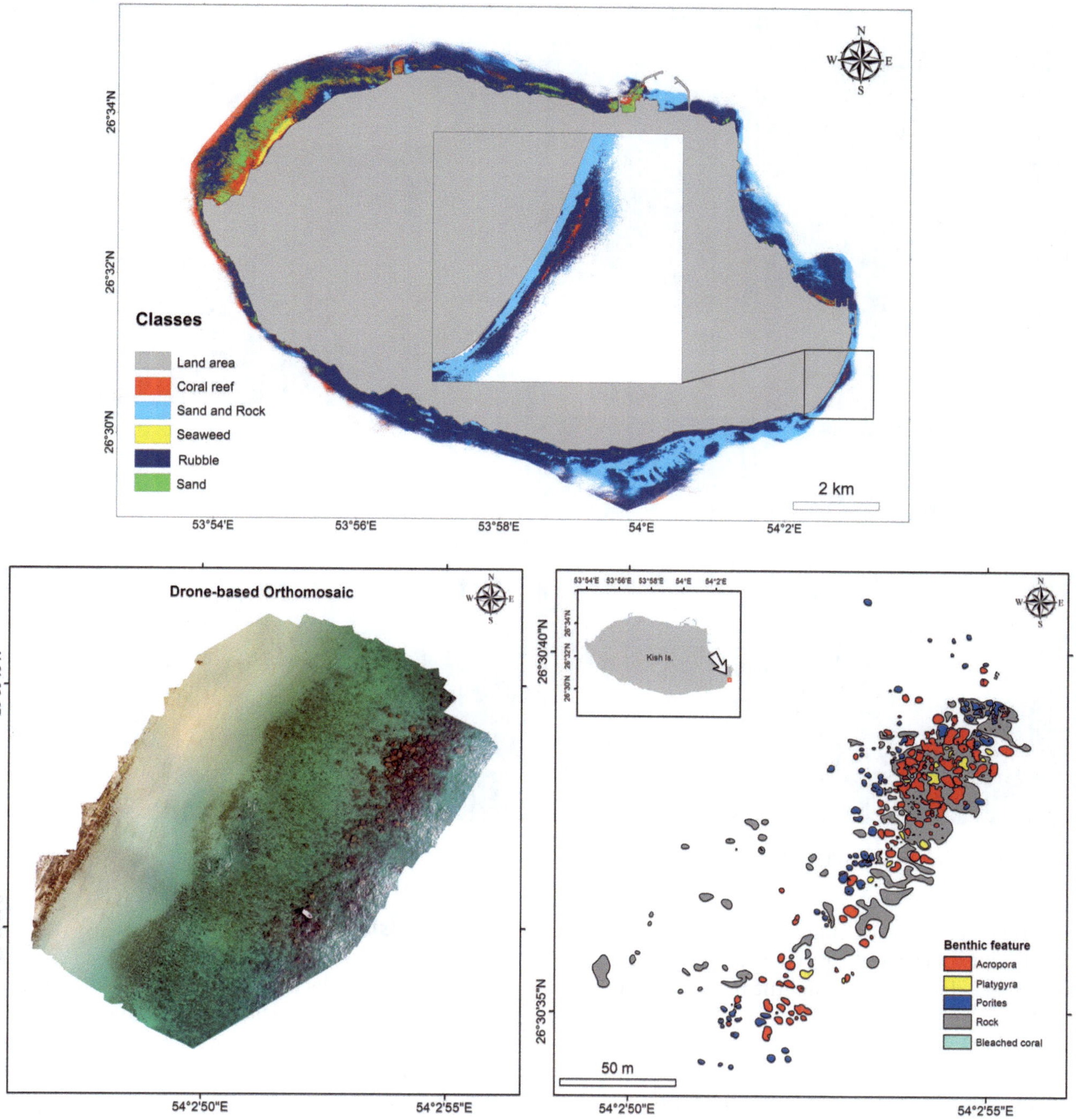

Fig. 2.12 Detailed map of substrate classes around Kish Island

2.4.10 Farur Island

The western and northern coasts of Farur Island feature significant coral reef formations, while the southern and eastern parts are mainly characterized by sand and rubble, with minimal coral presence (Fig. 2.13). Although designated as a protected area, there are no specific restrictions on marine life removal beyond generally applicable national or subnational regulations. The area of live coral begins at a depth of 2 meters and extends to a maximum depth of 15 meters. This area is primarily composed of finger corals from the genus *Porites*, *Platygyra* brain corals, and star corals from the genus *Dipsastraea* and *Favites*. Branching corals from the genus *Acropora* are also present, generally found at depths of 2–15 meters. Due to its offshore location and status as a military restricted zone, access to the reefs on Farur Island is limited, which has helped maintain

Plate 2.11 Underwater views of coral reefs in Kish Island

their relatively pristine condition. However, successive bleaching events have adversely affected the island's coral reefs, leading to a significant loss of *Acropora* corals.

2.4.11 Bani Farur (Lesser Faruar) Island

Coral reefs on Bani Farur Island are predominantly distributed along the northern and southern coastlines, with the eastern region characterized mainly by sand and rubble substrates (Fig. 2.14). Massive corals of the genus *Porites* are the most abundant, followed by brain corals from the genus *Platygyra*, as well as star corals of the genera *Dipsastraea* and *Favites*, while *Acropora* corals are comparatively scarce. The offshore location of Bani Farur Island restricts easy access to its coral reefs, aiding in their preservation and contributing to a relatively pristine state. However, recurrent bleaching events have impacted these reefs, causing a marked decline in *Acropora* populations.

2.4.12 Siri Island

Coral reefs on Siri Island encompass the island's perimeter, with higher densities along the northern and western coasts, whereas the southern and eastern regions are mainly composed of rock and rubble (Fig. 2.15). The eastern side descends steeply to the escarpment bank/shelf and ultimately to deeper waters, limiting coral reef development in this area. Conversely, the western, northern, and southern regions feature bank/shelf and reef fore structures that create favorable conditions for coral reef formation. The area around Siri Island has moderate coral cover due to the favorable water currents. Despite the moderate cover, the overall health of the reefs is relatively good. Dominant species include massive corals of the genus *Porites*, followed by brain corals from the genus *Platygyra*, star corals from the genera *Dipsastraea* and *Favites*, and branching *Acropora* corals.

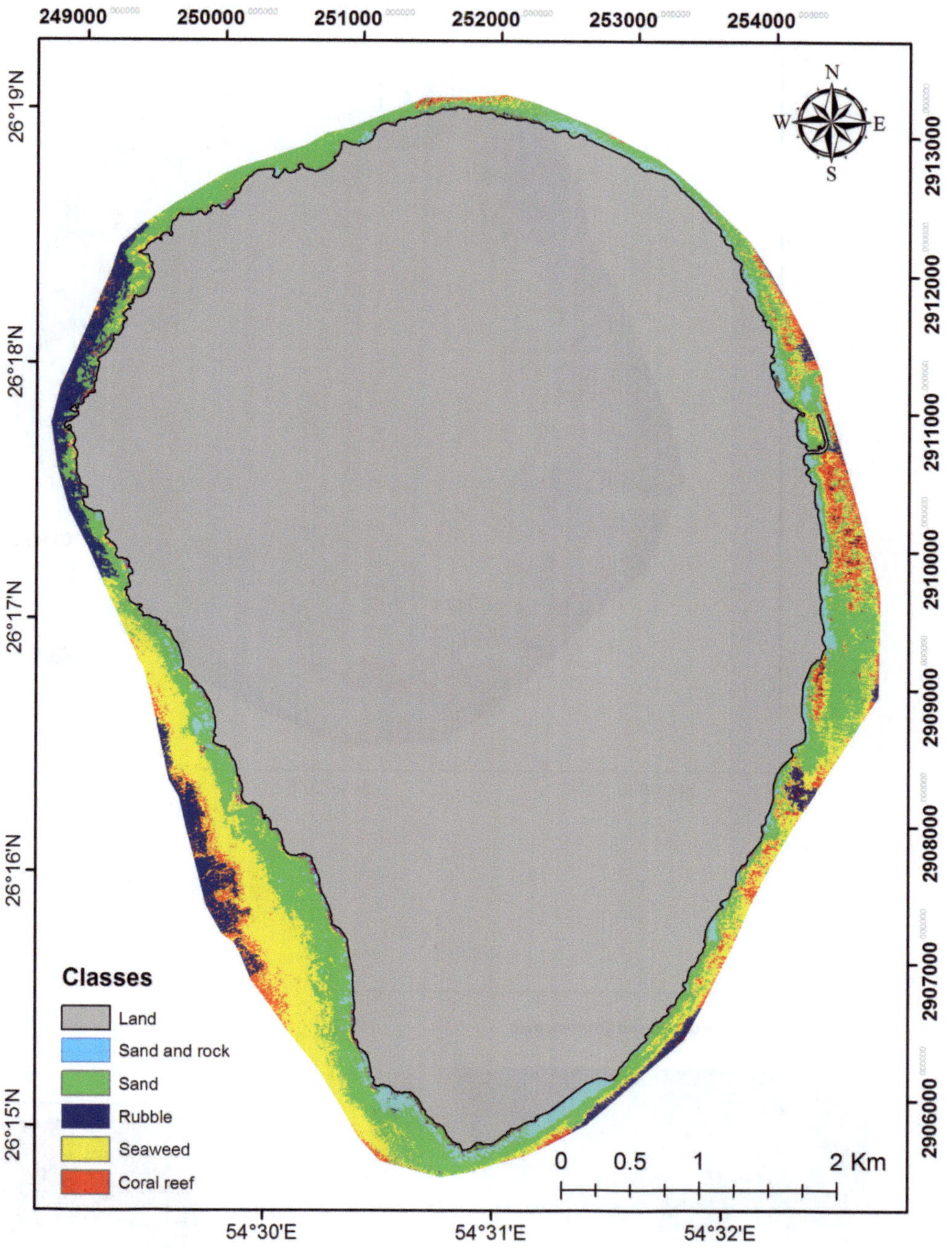

Fig. 2.13 Detailed map of substrate classes around Farur Island

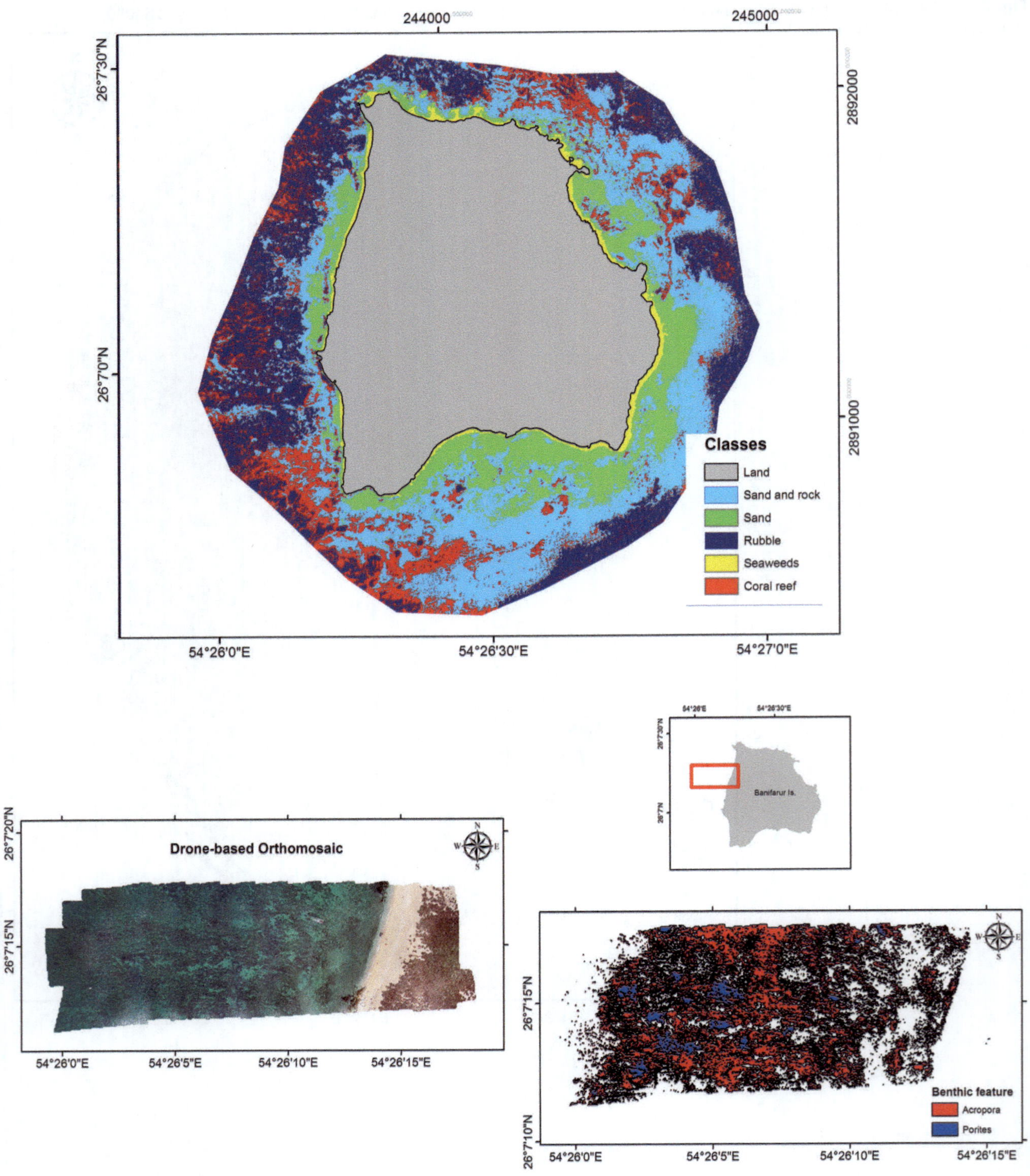

Fig. 2.14 Detailed map of substrate classes around Bani Farur Island

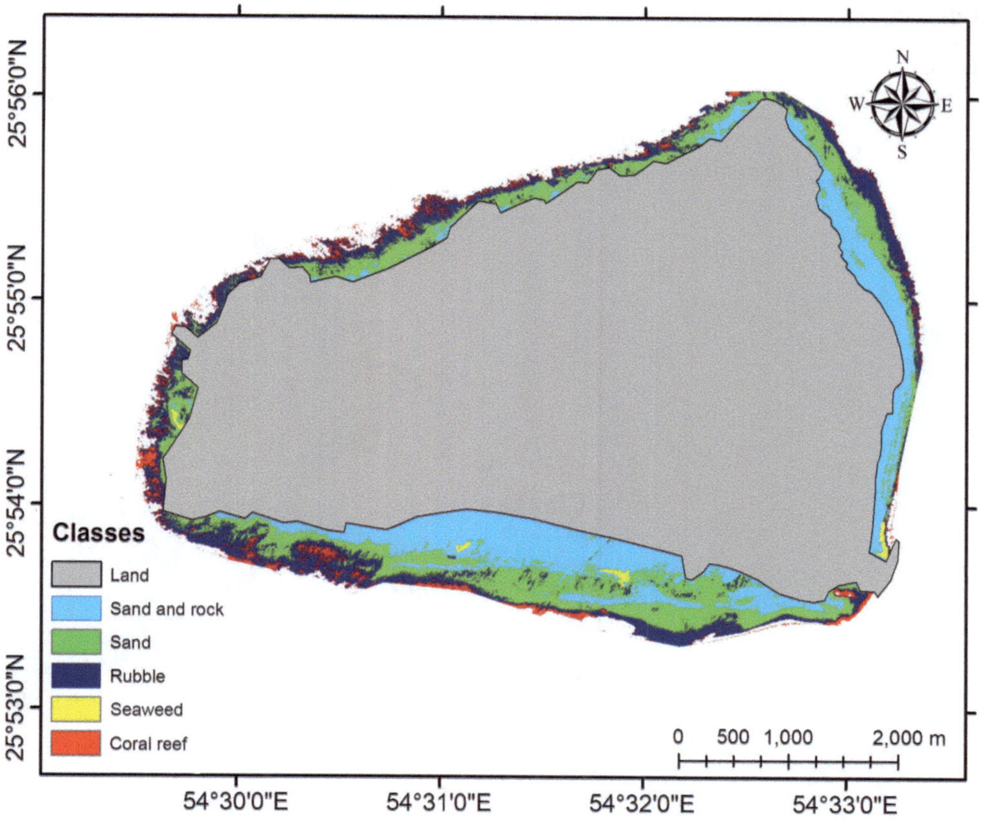

Fig. 2.15 Detailed map of substrate classes around Siri Island

2.4.13 Abu Musa Island

Abumusa Island is encircled by coral reefs, especially along its northern and western shores (Fig. 2.16). The eastern part of the island is predominantly composed of rubble and rock, with a sparse coral presence. The eastern side descends steeply to the escarpment bank/shelf and eventually to deep waters, limiting the formation of coral colonies. In contrast, the western, northwestern, and southwestern parts of the island feature bank/shelf and reef fore structures, creating suitable conditions for coral bed formation. The area surrounding Abu Musa Island has a high coral cover, attributed to the favorable water currents around the island. Despite the high coral cover, the overall health of the reefs is relatively low. The predominant genera include branching *Acropora* corals, followed by *Stylopora*, massive *Porites*, brain corals from the genus *Platygyra*, and star corals from the genera *Dipsastraea* and *Favites* (Plate 2.12).

2.4.14 Lesser Tunb Island

Coral reefs on Lesser Tunb Island are primarily found along the northeastern and southwestern shores, with coral patches interspersed among rocky areas (Fig. 2.17). The southern and western coasts are mainly characterized by rubble and sand. The area surrounding Lesser Tunb Island has a high coral cover, attributed to the favorable water currents around the island. Despite the high coral cover, the overall health of the reefs is relatively low. The dominant coral genera include branching *Acropora*, followed by *Stylopora*, massive *Porites*, brain corals from the genus *Platygyra*, and star corals from the genera *Dipsastraea* and *Favites*.

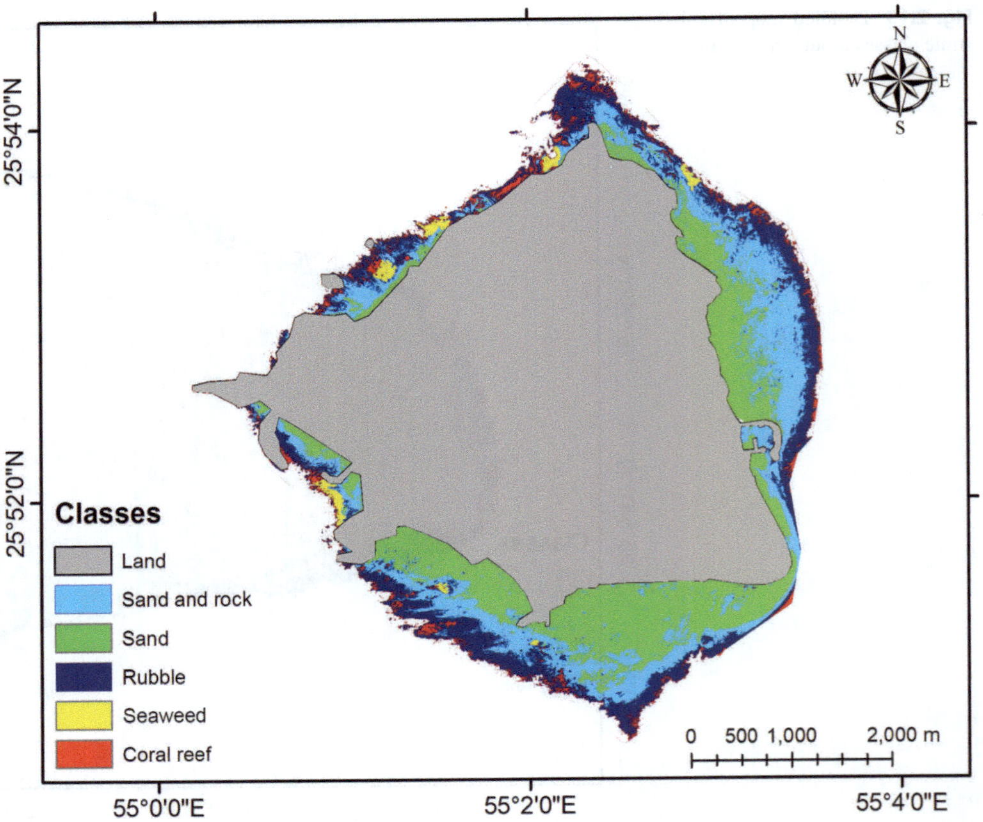

Fig. 2.16 Detailed map of substrate classes around Abu Musa Island

Plate 2.12 Underwater views of coral reefs in Abu Musa Island

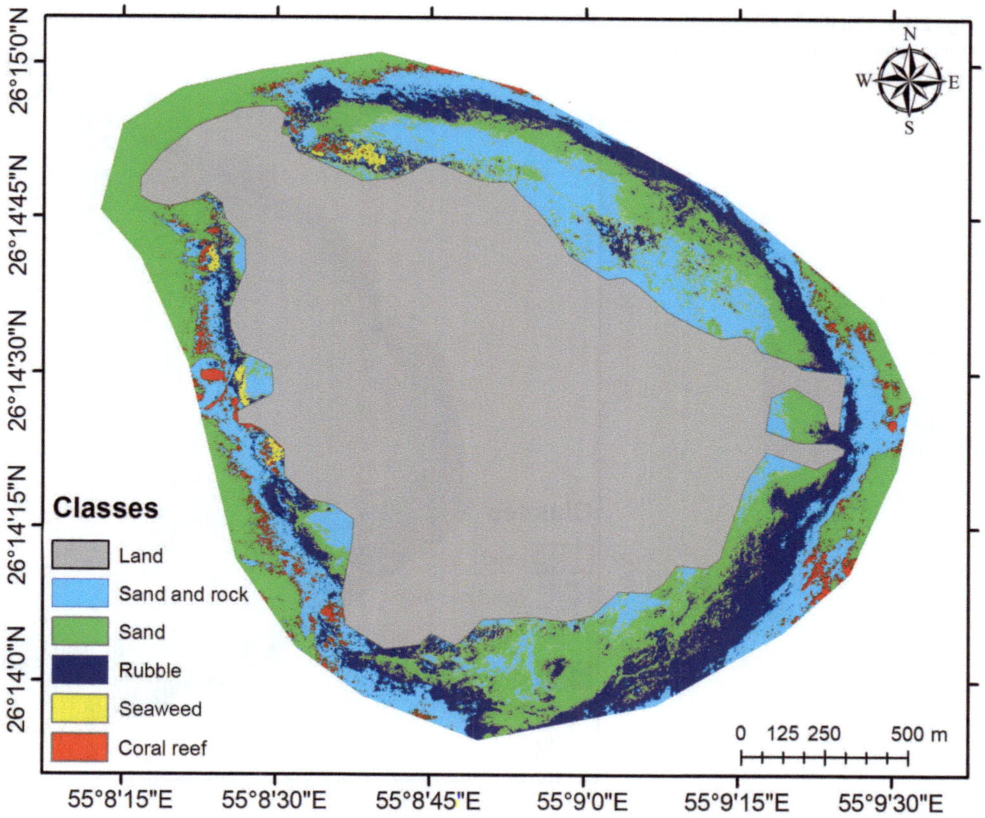

Fig. 2.17 Detailed map of substrate classes around Lesser Tunb Island

2.4.15 Greater Tunb Island

Coral reefs on Greater Tunb Island are primarily found along the northern shore, while the rest of the island's perimeter consists of rubble and rocky areas (Fig. 2.18). The southern coast is predominantly characterized by rocky substrates with minimal coral presence. Despite the low coral cover surrounding Greater Tunb Island, the corals are in good health. The predominant genera include branching *Acropora*, *Stylopora*, and *Montipora*, as well as massive *Porites*, brain corals from the genus *Platygyra*, and star corals from the genera *Dipsastraea* and *Favites*.

2.4.16 Hengam Island

Coral reefs on Hengam Island are predominantly located along the eastern and northern coasts, while the southern and western shores are mainly characterized by rubble and sand, with fewer coral patches (Fig. 2.19). The corals of Hengam Island are directly exposed to oceanic currents entering through the Strait of Hormuz, bringing cooler, less saline, clearer, and nutrient-rich waters. These currents create favorable conditions for a diverse range of coral species. The coral fauna is dominated by branching *Acropora* corals, followed by *Porites* and *Platygyra* (Plate 2.13). Areas with higher coral coverage have lower diversity and are composed of

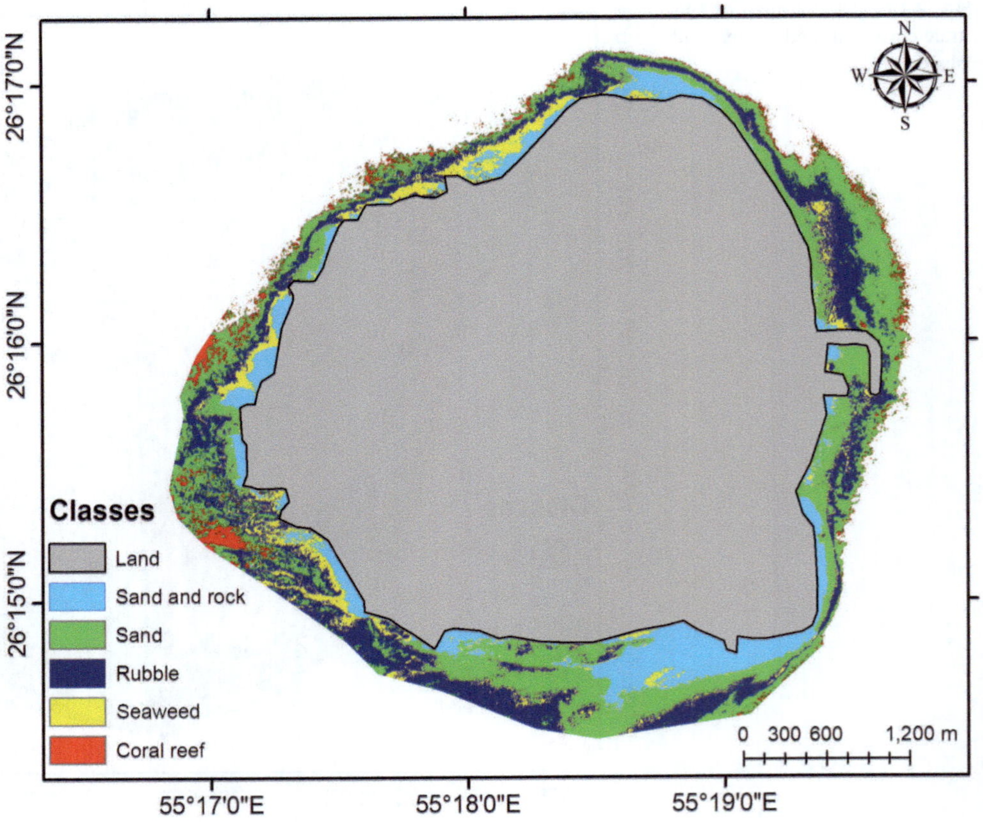

Fig. 2.18 Detailed map of substrate classes around Greater Tunb Island

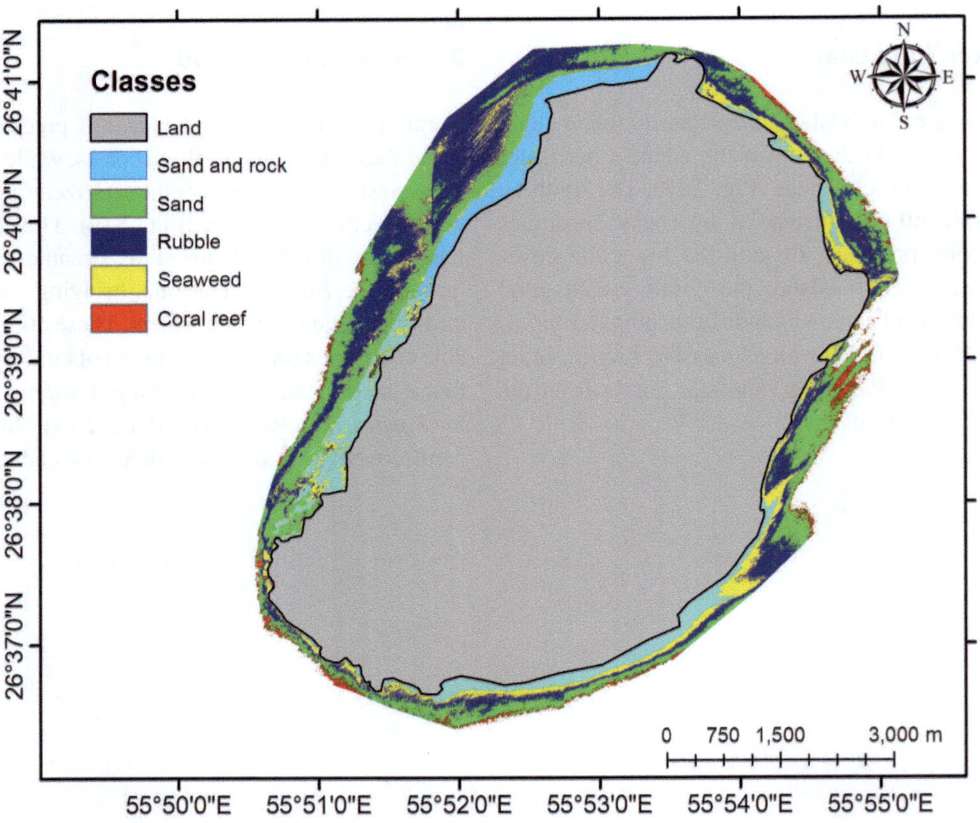

Fig. 2.19 Detailed map of substrate classes around Hengam Island

Plate 2.13 Underwater views of coral reefs in Hengam Island

mono-species corals. The rapid expansion of tourism in nearby Qeshm Island has led to an influx of boats transferring tourists to the reefs around Hengam Island for diving, fishing, and dolphin watching. Significant damage to the corals has been observed, primarily due to removal of corals, as well as damage from moored boats.

2.4.17 Larak Island

Coral reefs on Larak Island are primarily found along the northern and northeastern shores, with the southern coastline characterized by a mixture of rubble and sand and exhibiting sparse coral distribution (Fig. 2.20). The corals are directly exposed to oceanic currents entering through the Strait of Hormuz, which bring cooler, less saline, clearer, and nutrient-rich waters, creating favorable conditions for a diverse range of coral species. Consequently, coral reefs on Larak Island are characterized by both high species diversity and significant coverage. The coral fauna is dominated by branching *Acropora* corals, followed by large *Porites* and *Platygyra* (Plate 2.14). Generally, Larak Island has greater coral diversity and a relatively large coverage of branching corals compared to other islands. The principal threats to these reefs include ecotourism and coastal construction activities.

2.4.18 Qeshm Island

Coral reefs on Qeshm Island are dispersed along the northern coastline, especially in the northeastern regions (Fig. 2.21). The southern coast is predominantly rocky, with minimal coral cover. The coral fauna is dominated by *Porites* and *Platygyra*, followed by branching *Acropora* corals (Plate 2.15). Additionally, intertidal tide pools, particu-

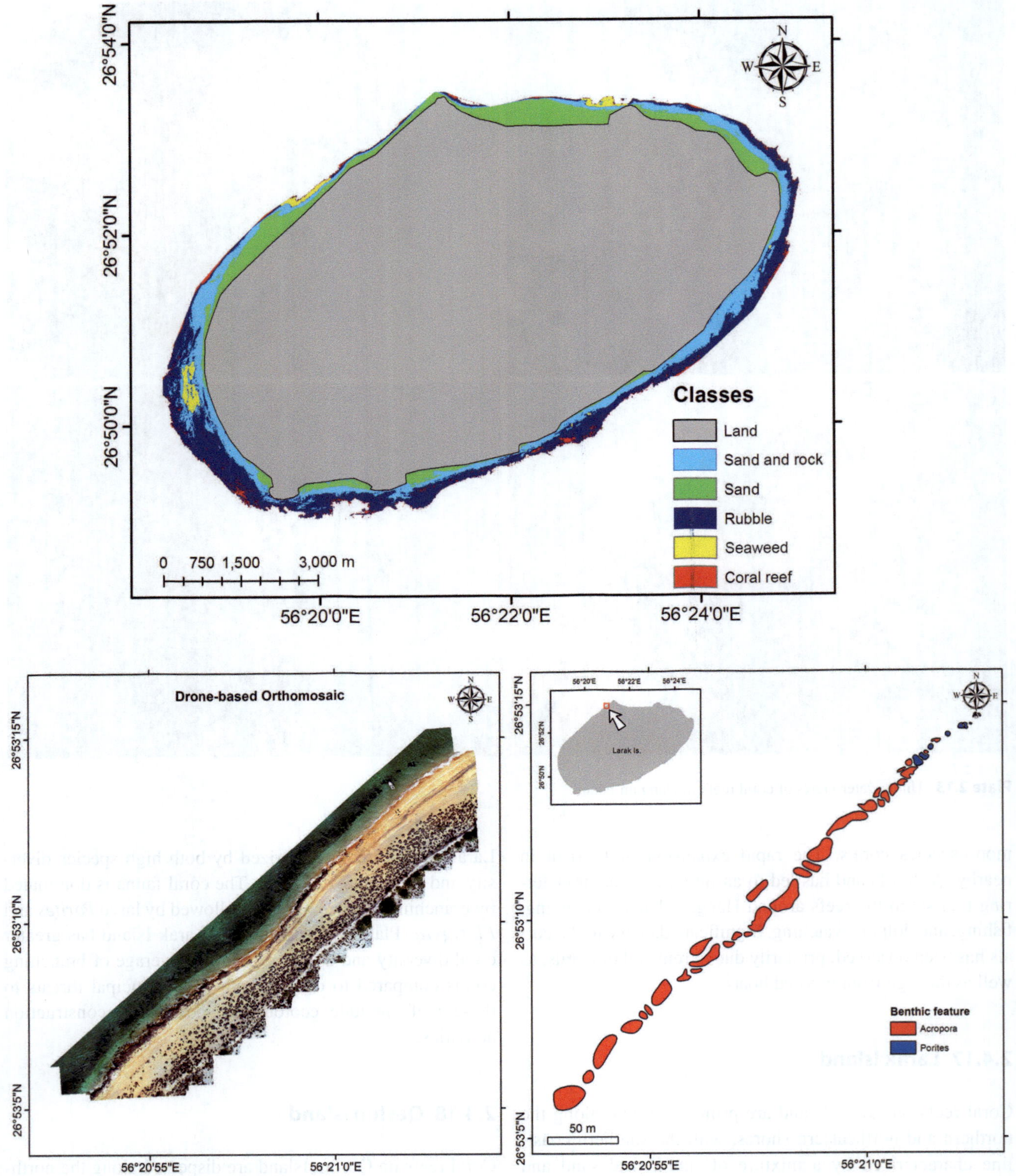

Fig. 2.20 Detailed map of substrate classes around Larak Island

Plate 2.14 Underwater views of coral reefs in Larak Island

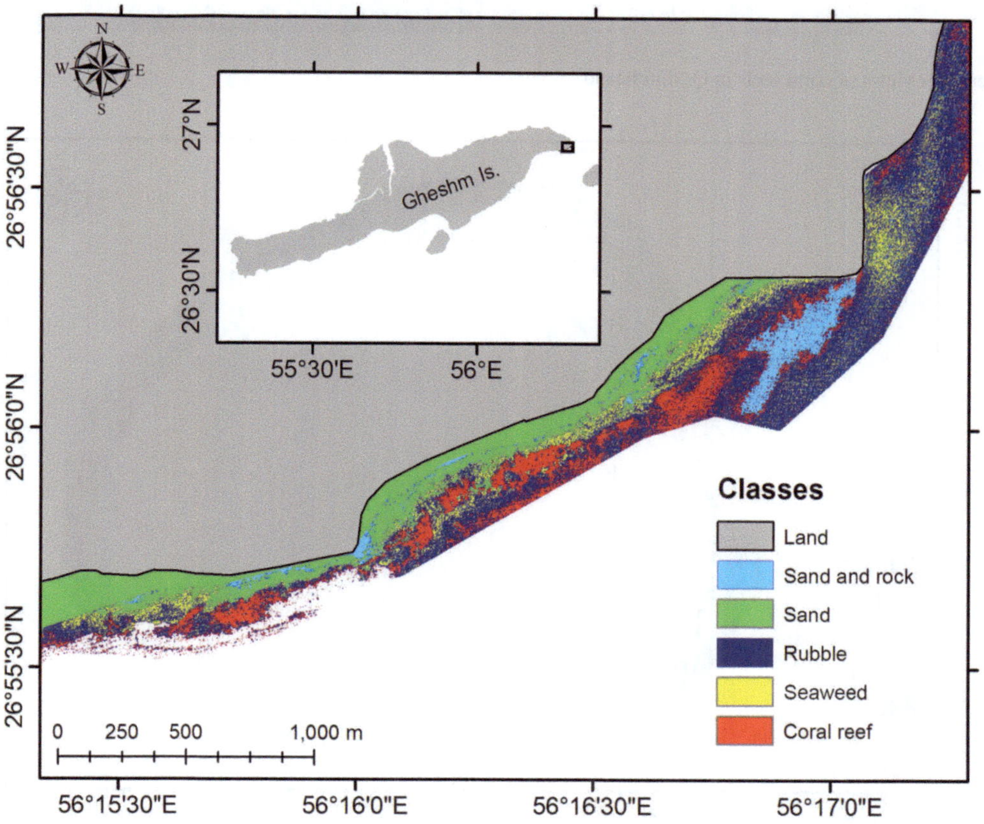

Fig. 2.21 Detailed map of substrate classes around Qeshm Island

Plate 2.15 Underwater views of coral reefs in Qeshm Island

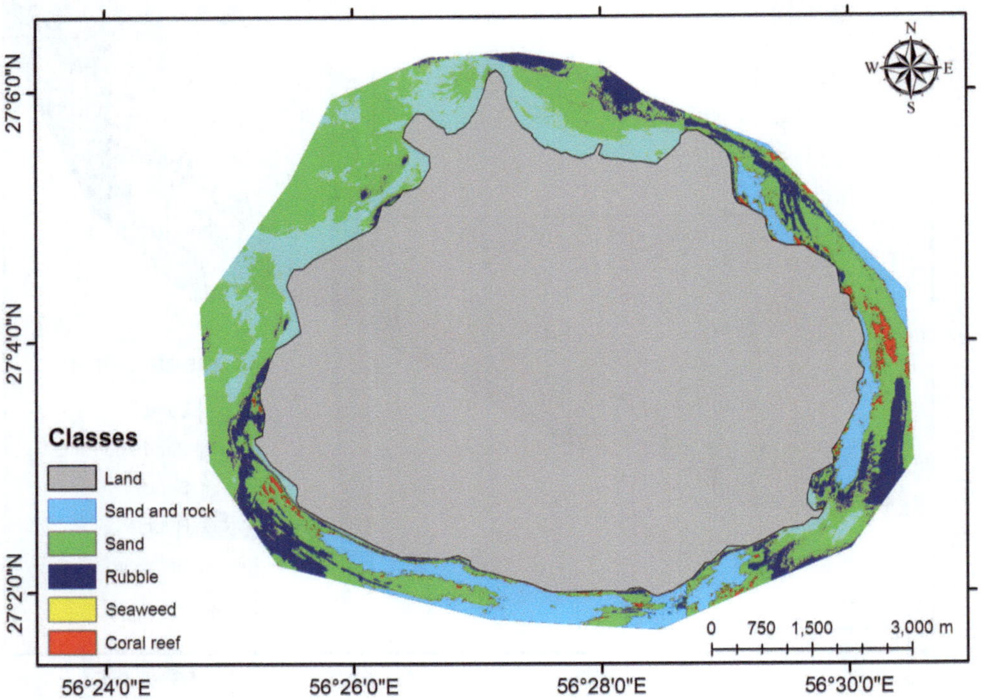

Fig. 2.22 Detailed map of substrate classes around Hormuz Island

larly on the southeastern coast, support diverse marine life, where hard corals and zoanthids coexist. These coral reefs are threatened by the expansion of intensive ecotourism and coastal developments.

2.4.19 Hormuz Island

On Hormuz Island, coral reefs are primarily distributed along the southern, southeastern, and western coasts, with fewer reefs on the northern shore due to prevalent rocky substrates (Fig. 2.22). The southeastern coast has the highest hard coral species diversity and the best reef development, coinciding with areas of optimal water clarity. Coral communities occur from the low water spring tide level down to depths of 4–5 meters, typically forming narrow bands of 10–15 meters, except on the western side, where they extend up to 50 meters offshore. Species diversity is highest in the southeast, gradually decreasing toward the north and west. Variations include a hypersaline stream impacting the western coral patch and significant wave exposure in the southeastern patch. Hematite-contaminated sediments from a nearby mine limit coral diversity in the southwestern area. Optimal coral growth on the western patch is attributed to clear, sheltered waters protected by a headland, while inflowing currents from the Gulf of Oman enhance water quality and nutrients along the southern and southeastern shores. In contrast, the northern waters are warmer, occasionally more saline, and turbid due to coastal influence and mangrove detritus. The coral fauna is dominated by *Porites* and *Platygyra* corals, with limited *Acropora* branching corals. Additionally, intertidal tide pools, particularly on the western side, support diverse marine life, where hard corals and zoanthids coexist.

2.5 Iran's Coral Growing Areas in the Gulf of Oman

2.5.1 Pozm Bay

Pozm Bay, located on the westernmost coast of Iran along the Gulf of Oman, contains coral reefs that form as dispersed, patchy formations primarily on its eastern side (Fig. 2.23). Positioned near the diverse coral ecosystems of the Indian Ocean, particularly those in India, the corals in Pozm Bay represent a subset of this broader faunal region. Dominant coral genera in the area include *Turbinaria*, *Porites*, *Platygyra*, *Dipsastraea*, and *Cyphastrea*. Unlike the coral reefs in Iran's Persian Gulf, the coral assemblages in Pozm Bay experience cooler environmental conditions due to monsoonal upwelling, which has shielded them from the severe bleaching events of 1996, 1998, and 2017 that heavily impacted Persian Gulf corals. However, high water turbidity from monsoonal upwelling restricts the development of these reefs, preventing them from forming continuous patch formations.

2.5.2 Chabahar Bay

Chabahar Bay, situated on the westernmost coast of Iran along the Gulf of Oman, hosts patchy coral reefs, predominantly distributed along its eastern coast (Fig. 2.23). Positioned near the diverse coral ecosystems of the Indian Ocean, especially those in India, the coral reefs of Chabahar Bay represent a subset of the broader coral diversity of the region. Key coral genera in this area include *Pocillopora*, *Stylopora*, *Acropora*, *Turbinaria*, *Porites*, *Platygyra*, *Dipsastraea*, and *Cyphastrea* (Plate 2.16). Unlike the coral reefs in the Persian Gulf, the coral assemblages in Chabahar Bay experience relatively cooler environmental conditions due to monsoonal upwelling, which has allowed them to avoid the severe bleaching events of 1996, 1998, and 2017 that greatly impacted Persian Gulf corals. However, the development and ecological health of the Chabahar Bay reefs have been substantially compromised by the expansion of Shahid Beheshti Port, as well as by the increased turbidity resulting from monsoonal upwelling, dredging, and port construction. In the southeastern region of Chabahar Bay, near the Ports and Maritime Organization's fuel storage facilities and adjacent to Shahid Beheshti Pier, an area covering more than five hectares was formerly populated by patchy coral reefs. In 2010, as part of the Shahid Beheshti Port development, these coral reefs were relocated to the Lipar Coral Site in the mid-eastern bay. Nonetheless, turbidity from dredging activities and the discharge of ballast water from ships continue to pose significant threats to the coral reefs' health in Chabahar Bay.

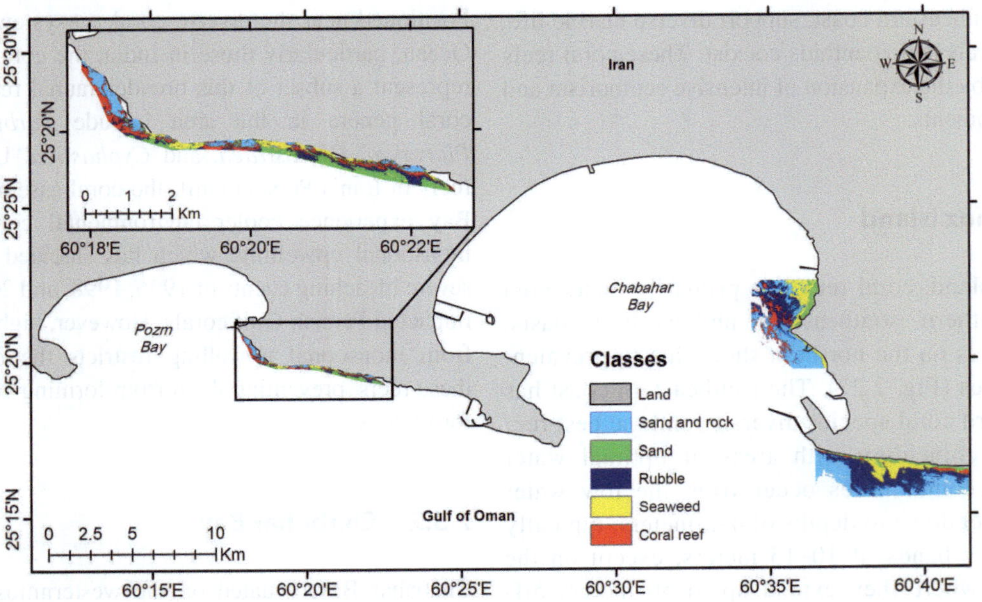

Fig. 2.23 Detailed map of substrate classes around Pozm and Chabahar bays

Plate 2.16 Underwater views of coral reefs in Chabahar Bay

References

Brewer PG, Dyrssen D (1985) Chemical oceanography of the Persian Gulf. Prog Oceanogr 14:41–55. https://doi.org/10.1016/0079-6611(85)90004-7

Coles SL (2003) Coral species diversity and environmental factors in the Arabian Gulf and the Gulf of Oman: a comparison to the Indo-Pacific region. Atoll Res Bull:1–19. https://doi.org/10.5479/si.00775630.507.1

Elliott AJ, Savidge G (1990) Some features of the upwelling off Oman. J Mar Res 48:319–333

Kabiri K, Pradhan B, Samimi-Namin K, Moradi M (2013) Detecting coral bleaching, using QuickBird multi-temporal data: a feasibility study at Kish Island, the Persian Gulf. Estuar Coast Shelf Sci 117:273–281. https://doi.org/10.1016/j.ecss.2012.12.006

Kabiri K, Rezai H, Moradi M (2018) Mapping of the corals around Hendorabi Island (Persian Gulf), using WorldView-2 standard imagery coupled with field observations. Mar Pollut Bull 129:266–274. https://doi.org/10.1016/j.marpolbul.2018.02.045

Kabiri K, Rezai H, Moradi M (2020) A drone-based method for mapping the coral reefs in the shallow coastal waters – case study: Kish Island, Persian Gulf. Earth Sci Informatics. https://doi.org/10.1007/s12145-020-00507-z

Kassler P (1973) The structural and geomorphic evolution of the Persian Gulf. In: Purser BH (ed) The Persian Gulf: Holocene carbonate sedimentation and diagenesis in a shallow epicontinental sea. Springer, Berlin, Heidelberg, pp 11–32

Lyzenga D (1978) Passive remote sensing techniques for mapping water depth and bottom features. Appl Opt 17:379–383. https://doi.org/10.1364/AO.17.000379

Morrison J (1997) Inter-monsoonal changes in the T-S properties of the near-surface waters of the Northern Arabian Sea. Geophys Res Lett Geophys RES LETT 24:2553–2556. https://doi.org/10.1029/97GL01876

Mumby PJ, Edwards AJ (2000) Water column correction approaches. In: Green EP, Mumby PJ, Edwards AJ, Clark CD (eds) Remote sensing handbook for tropical coastal management. Unesco, Paris, p 316

Pirazzoli PA, Reyss J-L, Fontugne M, Haghipour A, Hilgers A, Kasper HU, Nazari H, Preusser F, Radtke U (2004) Quaternary coral-reef terraces from Kish and Qeshm Islands, Persian Gulf: new radiometric ages and tectonic implications. Quat Int 120:15–27. https://doi.org/10.1016/j.quaint.2004.01.003

Reynolds RM (1993) Physical oceanography of the Persian Gulf, Strait of Hormuz, and the Gulf of Oman – results from the Mt Mitchell expedition. Mar Pollut Bull 27:35–59

Rezai H, Wilson S, Claereboudt M, Riegl B (2004) Coral reef status in the ROPME sea area: Arabian/Persian Gulf, Gulf of Oman and Arabian Sea. Status of Coral Reefs of the World, pp 155–170.

Riegl BM, Purkis SJ (2012) Coral reefs of the Gulf: adaptation to climatic extremes in the World's Hottest Sea. In: Riegl BM, Purkis SJ (eds) BT – coral reefs of the Gulf: adaptation to climatic extremes. Springer, Dordrecht, pp 1–4

Riegl BM, Purkis SJ, Al-Cibahy AS, Abdel-Moati MA, Hoegh-Guldberg O (2011) Present limits to heat-adaptability in corals and population-level responses to climate extremes. PLoS One. https://doi.org/10.1371/journal.pone.0024802

Ross DA, Uchupi E, White RS (1986) The geology of the Persian Gulf-Gulf of Oman Region: a synthesis. Rev Geophys 24:537–556. https://doi.org/10.1029/RG024i003p00537

Schils T, Wilson SC (2006) Temperature threshold as a biogeographic barrier in northern Indian Ocean macroalgae. J Phycol 42:749–756. https://doi.org/10.1111/j.1529-8817.2006.00242.x

Seibold E, Diester L, Fütterer D, Lange H, Müller P, Werner F (1973) Holocene sediments and sedimentary processes in the Iranian part of the Persian Gulf. In: Purser BH (ed) The Persian Gulf: Holocene carbonate sedimentation and diagenesis in a shallow epicontinental sea. Springer, Heidelberg, pp 57–80

Spalding MD, Ravilious C, Green EP (2001) World atlas of coral reefs, 1st edn. UNEP World Conservation Monitoring Centre. University of California Press, Berkley, USA

3 Scleractinian Corals of Iran

The coral reefs of the Persian Gulf and Gulf of Oman represent a considerably reduced subset of Indo-Pacific coral fauna. Only around 10% of the Indo-Pacific species are found in the Persian Gulf, with the community composition significantly altered from the typical Indo-Pacific assemblages (Coles 2003). The estimated numbers of coral genera in the Persian Gulf and Gulf of Oman are 28 and 33, respectively (Veron 1986). By contrast, the Red Sea is home to over 75 coral genera, including 57 reef-forming genera (Sheppard and Sheppard 1991). Of the 656 species among 109 genera of reef-building corals in the Indo-Pacific, about 10%, or 68 species across 28 genera, are found in the Persian Gulf, and 68 species across 33 genera are found in the Gulf of Oman (Coles 2003). Other studies have reported even fewer scleractinian coral species, with a maximum of 40 species identified in the Gulf (Rezai et al. 2004; Riegl and Purkis 2012b). It is believed that a detailed taxonomic reanalysis would further reduce the number of reliably recorded species to around 40 (Riegl et al. 2012). Despite harsh conditions, the Gulf is home to approximately 40 species of reef-building corals (Coles 2003) and 31 Alcyonacean corals (soft corals) (Samimi-Namin and Ofwegen 2009), representing a depleted yet characteristic fragment of the Indo-Pacific fauna (Purkis and Riegl 2012).

In both the Gulf of Oman and the Arabian Sea, reef-building potential remains low due to high rates of bioerosion driven by primary production from the Arabian upwelling system. The coral fauna in these regions is a subset of the Indo-Pacific fauna, intermixed with regional endemics. Faviids are particularly well represented, while acroporiids and fungiids are notably under-represented. At least 10 Southern Arabian endemic species have been identified, including two endemic genera, *Parasimplastrea* and *Calathiscus*, with the taxonomic positions of several other species yet to be confirmed (Rezai et al. 2004).

In the Persian Gulf, the highest number of species, with a total of 50, has been recorded from the offshore islands of Saudi Arabia (Basson et al. 1977; Burchard 1979), though this figure was suspected to be an overestimation (Vogt 1996). Following this, the UAE reported 34 scleractinian species (Riegl 1999). Iran boasts a rich coral fauna, likely the most diverse in the Gulf, attributed to more favorable oceanographic conditions (Riegl et al. 2012). The local species richness in the Gulf is subject to temporal fluctuations caused by mass mortality events that particularly impact branching *Acropora* (Shinn 1976; Riegl 1999). Three species are restricted to the Persian Gulf and nearby Arabian and Red Sea: *Acropora arabensis* (Hodgson and Carpenter 1995), *Acropora downingi* Wallace, 1999, and the endemic *Porites harrisoni* Veron, 2000 (Riegl et al. 2012). The closest faunistic proximity to other Indo-Pacific reefs is naturally to the Gulf of Oman and the Red Sea, due to a shared paleoceanographic history of restriction during the last sea-level low stand and simultaneous flooding during the Holocene transgression (Sheppard and Sheppard 1991; Uchupi et al. 1996). While the Red Sea has marked endemism with 18 species, the Persian Gulf shares its entire species pool with the Indian Ocean (Hodgson and Carpenter 1995; Wallace 1999; Veron 2000).

The coral fauna in the Persian Gulf, comprising approximately 40 species, is as diverse as that found in the Caribbean (Chiappone et al. 1996). Similar to many other reef areas, the scleractinian genus *Acropora* plays a crucial role as the primary framework builder and dominates the shallowest regions. In slightly deeper waters, massive *faviids* typically prevail below the *Acropora*-dominated zone. The tabular *Acropora downingi* species group can establish well-developed, rigid frameworks. However, framework production has been significantly hindered by mass mortality events that have decimated large portions of the frame-builder populations. These coral mass mortalities are primarily caused by temperature anomalies, whether negative or positive (Coles 1988; Coles and Fadlallah 1991; Fadlallah et al. 1995; Riegl 2001, 2003; Grandcourt 2012). Among the non-*Acropora* frame builders, the genus *Porites* is particularly important. In the Persian Gulf, *Porites harrisoni* forms

M. R. Shokri, K. Kabiri, *Coral Reefs of Iran*, Coral Reefs of the World 22, https://doi.org/10.1007/978-3-032-00598-4_3

coherent, rigid frameworks. In regions with more extreme salinity and temperature variations, smaller frameworks of *Porites* corals replace the large *Acropora* frameworks (Riegl and Purkis 2012a).

Each summer, the Persian Gulf, the world's hottest sea, experiences sea surface temperatures (SSTs) that regularly exceed 35 °C, with coral communities persisting for several months at temperatures around 34 °C (Riegl et al. 2011). Despite these extreme conditions, coral communities thrive throughout the Gulf, exhibiting the highest known bleaching thresholds globally (Riegl et al. 2011). As a biogeographic subset of the Indian Ocean, the Persian Gulf has also been affected by increased SSTs. The Gulf is characterized by environmental extremes, with salinity levels frequently exceeding 45 ppt, and SSTs fluctuating annually from winter lows below 12 °C to summer highs above 36 °C (Coles and Fadlallah 1991; Sheppard et al. 1992). These harsh conditions are selective for corals adapted to such extremes, enabling their survival at temperatures that would typically be lethal elsewhere (Coles 2003). Consequently, the dominant taxa in Persian Gulf reefs differ from those in the Indo-Pacific, with the Gulf fauna predominantly comprising more tolerant taxa such as faviids and siderastreids, while the more sensitive acroporids are under-represented (Coles 2003).

The Iranian reefs in the Persian Gulf occur as fringing and patch reefs across various parts of the mainland coast, around islands, and on some submerged platforms. Mainland reefs are present in Chabahar and Pozm bays, both located in the northeastern most part of the Gulf of Oman. Unlike Iran's reefs in the Persian Gulf, the coral assemblages of Chabahar and Pozm bays experience the relatively cool environmental conditions associated with monsoonal upwelling. The most diverse Iranian reef exists around Larak Island in the Strait of Hormuz, hosting 40 hermatypic coral species (Vajed Samiei et al. 2013). Species richness in other reefs has been estimated at 21 species in Kharg and Khargu Islands, 14 in Nayband Bay, 29 in Farur Island, 26 in Sirri Island, 20 in Hendorabi Island, 28 in Kish Island, 23 in Qeshm Island, 17 in Hengam Island, and 17 in Chabahar Bay (Fatemi and Shokri 2001; Kavousi et al. 2011; Aminrad and Azini 2013; Moradi et al. 2014; Salimi et al. 2018). Historically, Iranian coral communities were dominated by *Acropora* corals at depths of 0–6 meters, where this species was particularly abundant (Harger 1984; Shokri et al. 2000; Shokri 2001; Wilson et al. 2002; Rezai et al. 2004). Since the 1996/1998 bleaching events, live coral cover, especially Acroporidae, has declined by over 90% in many areas, primarily due to multiple bleaching events caused by rising temperatures (Riegl et al. 2018). The 2017 summer bleaching event resulted in the loss of more than 95% of *Acropora* corals in all Iranian reefs in the Persian Gulf (Shokri and Mohammadi 2021). Corals in Chabahar Bay escaped the devastating impacts of the 1996, 1998, and 2017 bleaching events (Ghazilou 2019). However, they experienced a bleaching event in the summer of 2018 when the water temperature rose above 30 °C, resulting in near 100% bleaching of massive corals such as *Dipsastraea* and *Cyphastrea*, about 10–50% bleaching of *Acropora* corals, and less than 5% bleaching of *Pocillopora* corals (Ghazilou 2019).

The Iranian reefs have been severely damaged due to dredging and port construction around the reefs in Assaluyeh Port near Nayband Bay, Kish Island, Hendorabi Island, and Chabahar Bay. The adverse impacts of these activities have included the direct loss of corals caused by the removal or burial of reefs and lethal or sub-lethal stress to corals caused by elevated turbidity and sedimentation rates.

The following sections provide detailed descriptions of individual coral species found in Iran's reefs within the Persian Gulf and the Gulf of Oman. While significant efforts have been made to compile a comprehensive list of scleractinian corals, further research is required to confirm additional records. All species discussed in this text are found in relatively shallow waters (<20 m). For specimens with uncertain identification, the term "cf." is used between the genus and species names (e.g., *Acropora* cf. *microphthalma*). For specimens where the species remains unidentified but the genus is known, "sp." follows the genus name (e.g., *Acropora* sp.).

Underwater photographs of the observed coral colonies illustrate typical growth forms, while close-up images highlight polyp structure details. Additionally, photographs of calices provide essential information for taxonomic identification. For each coral species, details such as family, growth form, habitat type, abundance, average colony size, corallite diameter, and ecological notes are presented.

3.1 Glossary of Terms Utilized in Coral Species Identification

3.1.1 Colony Shape and Form

The coralla of hard corals exhibit various growth forms, which are instrumental in their identification and description. The following are photographs and definitions of some of the major growth forms observed among Iranian corals (Plate 3.1):

1. Branching: Describes a branch with compact radial sub-branches.

2. Massive: Colonies that are solid, typically hemispherical, or have approximately equal dimensions in all directions.
3. Sub-massive: Colonies that tend to form small columns, knobs, or wedges.
4. Foliose: Coral colonies that are attached at one or more points, resembling leaves or plates in appearance.
5. Encrusting: Coral colonies that form a thin layer or crust over the underlying substratum.
6. Solitary: Corals composed of single individuals.

1	2
3	4
5	6

Plate 3.1 Major growth forms observed among Iranian corals

3.1.2 Corallite Types

The organization of the corallites or polyps on the colonies is usually divided into a few general categories that help differentiate different families of corals. These categories highlight the diversity and complexity of coral structures, each adapted to their unique environments. Here's a brief summary of the different categories (Veron 2000):

	Flabellate: In flabellate or flabello-meandroid colonies, polyps form vertical, sheet-like structures. Distinct meanders are separated by intervening empty spaces. The only species with a flabello-meandroid growth form found in the Gulf of Oman is *Euphyllia divisa*.
	Phaceloid: Polyps rise individually and remain separated from neighboring polyps, resembling long tubes. This mode of development is relatively rare, with only a few species in Oman exhibiting this colony type, including *Blastomussa* and *Parasimplastrea*. During budding, a fork forms in the skeleton, allowing the two polyps to continue growing individually. This growth form is more easily observed in longitudinal slices of the colony.
	Plocoid: Polyps rise from the colonies in a manner resembling small knobs. Each corallite possesses its own wall and originates from an interpolyp platform. Numerous species exhibit a plocoid growth form, especially within the Merulinidae family, including genera such as *Dipsastraea*, *Echinopora*, and *Cyphastrea*.
	Cerioid: Cerioid colonies are characterized by polyps sharing common walls, which give the appearance of covering the entire coral surface. This growth form is typical of the genera *Favites* and *Goniopora*. The distinction between plocoid and cerioid growth forms can be quite subtle.
	Meandroid: Meandroid colonies are characterized by long valleys composed of multiple polyps that are interconnected. This growth form is typical of the genera *Platygyra* and *Symphyllia*, commonly referred to as brain corals.
	Dendroid: The polyps form a loosely structured, limited-branching tree. The non-reef building species within the genera *Tubastrea* and *Dendrophyllia* commonly exhibit a dendroid growth form and display diverse arrangements.
	Thamnasteroid: In thamnasteroid colonies, corallites exhibit confluent septa and lack distinct boundaries. This growth form is characteristic of the genus *Psammocora*.

3.2 Pictograms

3.2.1 IUCN Red List Code

Information on the distinguishing characteristics and habitats of each coral species is provided. Additionally, the conservation status of each species, as classified by the IUCN Red List, is presented through a color-coded strip as follows:

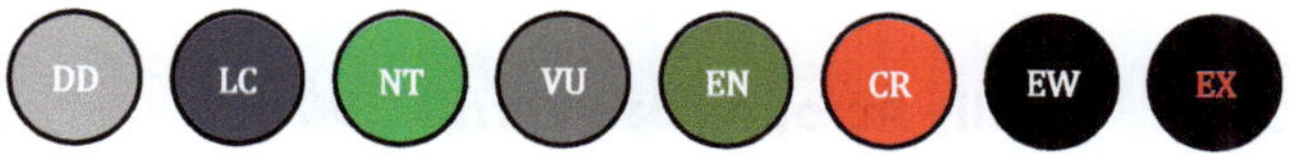

- Data Deficient (DD): A taxon is Data Deficient when there is inadequate information to make a direct, or indirect, assessment of its risk of extinction based on its distribution and/or population status. A taxon in this category may be well studied, and its biology well known, but appropriate data on abundance and/or distribution are lacking.
- Least Concern (LC): A taxon is Least Concern when it has been evaluated against the Red List criteria and does not qualify for Endangered Near Threatened.
- Near Threatened (NT): A taxon is Near Threatened when it has been evaluated against the criteria but does not qualify for Critically Endangered, Endangered or Vulnerable now, but is close to qualifying for or is likely to qualify for a threatened category in the near future.is close to qualifying for or is likely to qualify for a threatened category in the near future.
- Vulnerable (VU): A taxon is Vulnerable when the best available evidence indicates that it meets any of the criteria A to E for Vulnerable, and it is therefore considered to be facing a high risk of extinction in the wild.
- Endangered (EN): A taxon is Endangered when the best available evidence indicates that it meets any of the criteria A to E for Endangered, and it is therefore considered to be facing a very high risk of extinction in the wild.
- Critically Endangered (CR): A taxon is Critically Endangered when the best available evidence indicates that it meets any of the criteria A to E for Critically Endangered, and it is therefore considered to be facing an extremely high risk of extinction in the wild.
- Extinct in the Wild (EW): A taxon is Extinct in the Wild when it is known only to survive in cultivation, in captivity or as a naturalized population (or populations) well outside the past range. A taxon is presumed Extinct in the Wild when exhaustive surveys in known and/or expected habitat, at appropriate times (diurnal, seasonal, annual), throughout its historic range have failed to record an individual. Surveys should be over a time frame appropriate to the taxon's life cycle and life form.
- Extinct (EX): A taxon is Extinct when there is no reasonable doubt that the last individual has died. A taxon is presumed Extinct when exhaustive surveys in known and/or expected habitat, at appropriate times (diurnal, seasonal, annual), throughout its historic range have failed to record an individual. Surveys should be over a time frame appropriate to the taxon's life cycle and life form.

3.2.2 IUCN Population Trend

The current population trend, as assessed by the IUCN Red List website, is displayed. There are four possible descriptions for the population trend at the time of assessment: increasing, decreasing, stable, or unknown.

3.2.3 Levels of Abundance in Iran's Reefs

Indication of the possibility of encountering each individual coral species is presented under the heading 'Abundance' based on a 5-point DAFOR scale (DeVantier and Turak 2017) as followings:

Scale	Category	Abundance	Symbol
1	Rare	1–2 colonies	
2	Infrequent	3–5 colonies	
3	Frequent	6–20 colonies	
4	Common	21–50 colonies	
5	Dominant	51 + colonies	

This collection includes not only all the common coral species but also some uncommon and rare ones.

3.2.4 Population Trend in Iran's Reefs

The current population trend for each coral species, as assessed through long-term observations by the authors, is displayed. There are four possible descriptions for the population trend at the time of assessment: increasing, decreasing, stable, or unknown.

3.2.5 Distribution Map

The distribution map of individual coral species within the reefs of Iran, in the Persian Gulf and the Gulf of Oman, was compiled through personal observations and corroborated with data from published reports. The map employs various color-coded indicators: Dark Green circles denote confirmed presence, Green circles represent strongly predicted presence, Light Green circles indicate doubtful presence, and Pale Green circles signify the absence of any recorded presence of the respective coral species.

3.3 Scleractinian Species in Iran's Reefs

The coral species described in this book are based on specimens collected by the authors from Iran's reefs in the northern Persian Gulf and Gulf of Oman. Consequently, species previously recorded by other researchers but not encountered during our surveys are not included. It is likely that additional species will be identified and added to this collection in the future. Nonetheless, the described taxa represent the majority of scleractinian corals found in Iranian reefs across these regions.

Some of the identified species have an uncertain taxonomic status. While their family and genus classifications are likely accurate, species-level identifications may require further confirmation. For such cases, the Latin prefix **cf.** (from confer, meaning "compare with") is used before the species name to indicate resemblance to a particular species while leaving room for potential taxonomic revision.

Of the 18 extant families of scleractinian corals, 14 are represented in the reefs of northern Persian Gulf and Gulf of Oman. This book describes a total of 56 species from 31 genera, of which 51 species from 31 genera are reported from the Persian Gulf, and 30 species from 19 genera are recorded from the Gulf of Oman. Among these, three species are azooxanthellate. The described corals are organized by family, followed by genus.

The family classifications used in this book follow the system proposed by J.E.N. Veron (2000) in his comprehensive revision of scleractinian corals. However, family, genus and species names were updated according to the World Register of Marine Species (WoRMS Editorial Board 2004). As is common in taxonomy, these classifications remain subject to change and are increasingly informed by advancements in molecular genetics.

For each species, a brief description of key morphological characteristics is provided, accompanied by illustrations. While color is occasionally indicative, it is rarely diagnostic due to variations in appearance in photographs and under different underwater conditions. The distribution section outlines the species' prevalence and preferred habitats, though it is important to note that many corals are adaptable and occur in diverse environments.

3.3.1 Family Acroporidae Verrill, 1901

The family Acroporidae comprises colonial, hermatypic coral species. With the exception of *Astreopora*, members of this family are characterized by small corallites with septa typically exhibiting two or fewer cycles. The columellae are either absent or poorly developed. This family includes seven genera: *Acropora*, *Alveopora*, *Anacropora*, *Astreopora*, *Enigmopora*, *Isopora*, and *Montipora*. Of these, *Alveopora*, *Enigmopora*, and *Isopora* have not yet been recorded from Iranian reefs in the Persian Gulf and the Gulf of Oman.

3.3.1.1 Genus *Acropora* Oken 1815

With approximately 180 species (Veron 2000), *Acropora* is the largest and most widespread genus of extant scleractinian corals in tropical coral reefs. *Acropora* species are among the fastest-growing corals, with linear extension rates exceeding 20–30 cm per year in some cases (Yap and Gomez 1984; Crabbe and Smith 2005; Weil et al. 2020). These corals exhibit a branched or tabular morphology, with colony shapes ranging from bushy, plate-like, and ramose to arborescent, and are rarely encrusting or submassive (Fig. 3.1). The branches are characterized by the presence of both axial and radial corallites. Axial corallites guide the direction of branch growth, while radial corallites, which develop from axial corallites, influence branch thickness and overall appearance. Species within this genus are defined by their growth patterns, wherein axial corallites elongate and bud radial corallites at branch tips (Wells 1956; Wallace 1978; Veron et al. 1984). Radial corallites exhibit significant variability in shape (Fig. 3.2), a feature of taxonomic importance. Corallites lack columellae and typically have two cycles of septa. Both the corallite walls and coenosteum are porous. Key traits for species identification within this genus include the colony growth form, the structure of radial corallites, and the characteristics of the coenosteum.

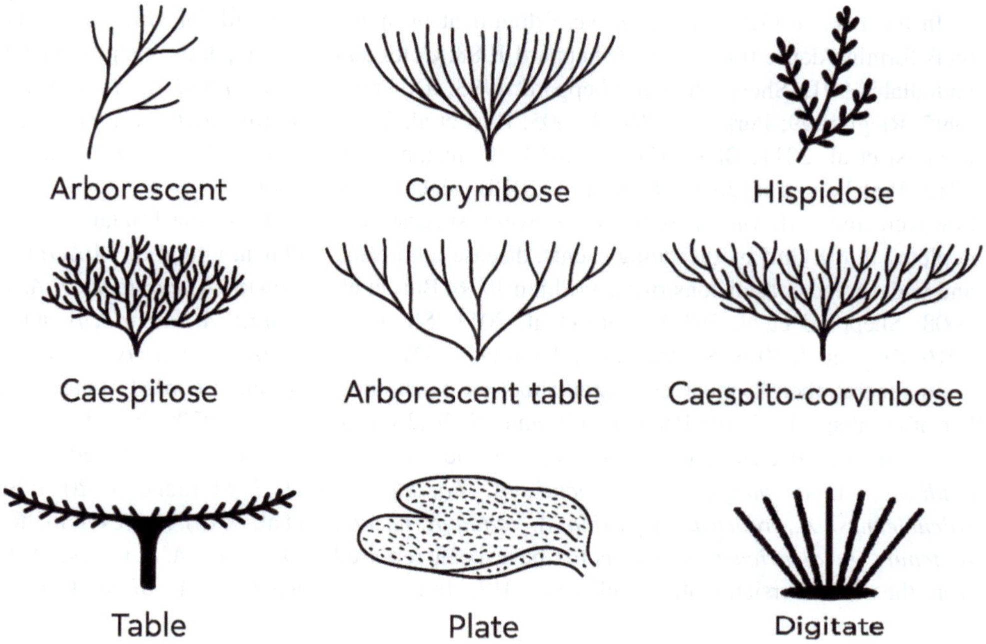

Fig. 3.1 Colony shapes in *Acropora* corals. (Adopted from Wallace and Dai 1997)

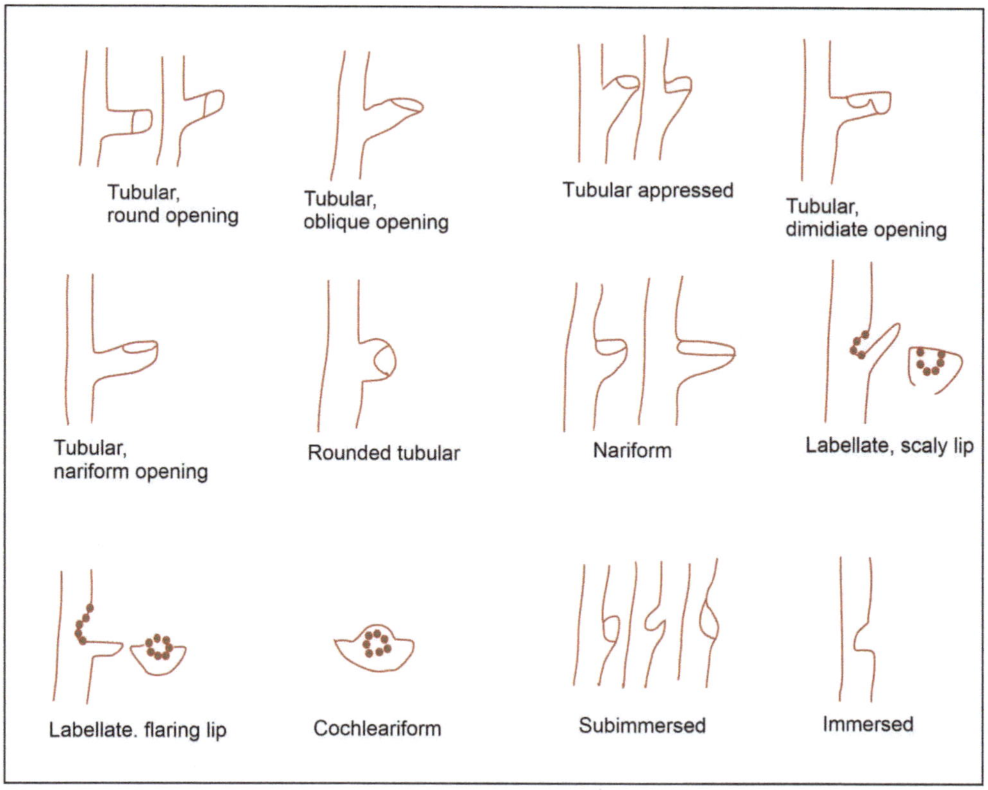

Fig. 3.2 Radial corallite shape categories in *Acropora* corals. (Adopted from Wallace and Dai 1997)

In the Persian Gulf, *Acropora* were dominant on many reefs forming dense frameworks (Sheppard 1988; Coles and Fadlallah 1991; Sheppard and Sheppard 1993; Downing 1995; Riegl 1999; Purkis and Riegl 2005; Burt et al. 2008; Kavousi et al. 2011; Bauman et al. 2013; Buchanan et al. 2016; Pratchett et al. 2016; Riegl et al. 2018). Since 1996; however, most *Acropora* populations were successively depleted in the Gulf by bleaching events, disease outbreaks, and widespread coastal construction (Shinn 1976; Burt et al. 2008; Sheppard et al. 2010b; Sale et al. 2011; Sheppard 2016; Riegl et al. 2018; Shokri and Mohammadi 2021).

In general, there is much confusion about the exact number of coral species in the Persian Gulf, and this is also true for *Acropora* species. Twelve *Acropora* species (i.e., *A. clathrate*, *A. downingi*, *A. pharaonis*, *A. horrida*, *A. valenciennesi*, *A. arabensis*, *A. florida*, *A. valida*, *A. nasuta*, *A. tenuis*, *A. divaricate*, *A. muricata*) have been reported from the entire Persian Gulf (Riegl et al. 2012). In contrast, the Gulf of Oman demonstrates a more limited diversity, encompassing approximately 11 species (i.e., *A. clathrate*, *A. downingi*, *A. pharaonis*, *A. khayranesis*, *A. gemmifera*, *A. paragemmifera*, *A. nasuta*, *A. cf microphthalma*, *A. valenciennesi*, *A. arabensis*, *A. valida*) (Coles 1996; Claereboudt 2006).

From the Iranian reefs in the Persian Gulf and Gulf of Oman, a total of 14 *Acropora* species (i.e., *A. arabensis*, *A. aspera*, *A. cervicornis*, *A. clathrate*, *A. downingi*, *A. gemmifera*, *A. horrida*, *A. microphthalma*, *A. mossambica*, *A. muricata*, *A. nausta*, *A. pharaonis*, *A. tortuosa*, *A. valida*) are so far recorded (Fatemi and Shokri 2001; Maghsoudlou 2008, 2020; Sadeghi and Loghmani 2009; Rahmani and Rahimian 2013; Vajed Samiei et al. 2013; Aminrad and Azini 2013; Moradi et al. 2014; Sharif Ranjabr et al. 2015; Salimi et al. 2018). However, in this book only seven species (i.e., *A. arabensis*, *A. downingi*, *A. hyacinthus*, *A. microphthalma*, *A. muricata*, *A. valida*, *Acropora* sp.) were recorded.

Acropora arabensis (Hodgson and Carpenter 1995)

Arabian staghorn coral			
IUCN red list assessment		Status in Iran's reefs	
Status	Population trend	Abundance	Population trend
NT	Decreasing		Decreasing

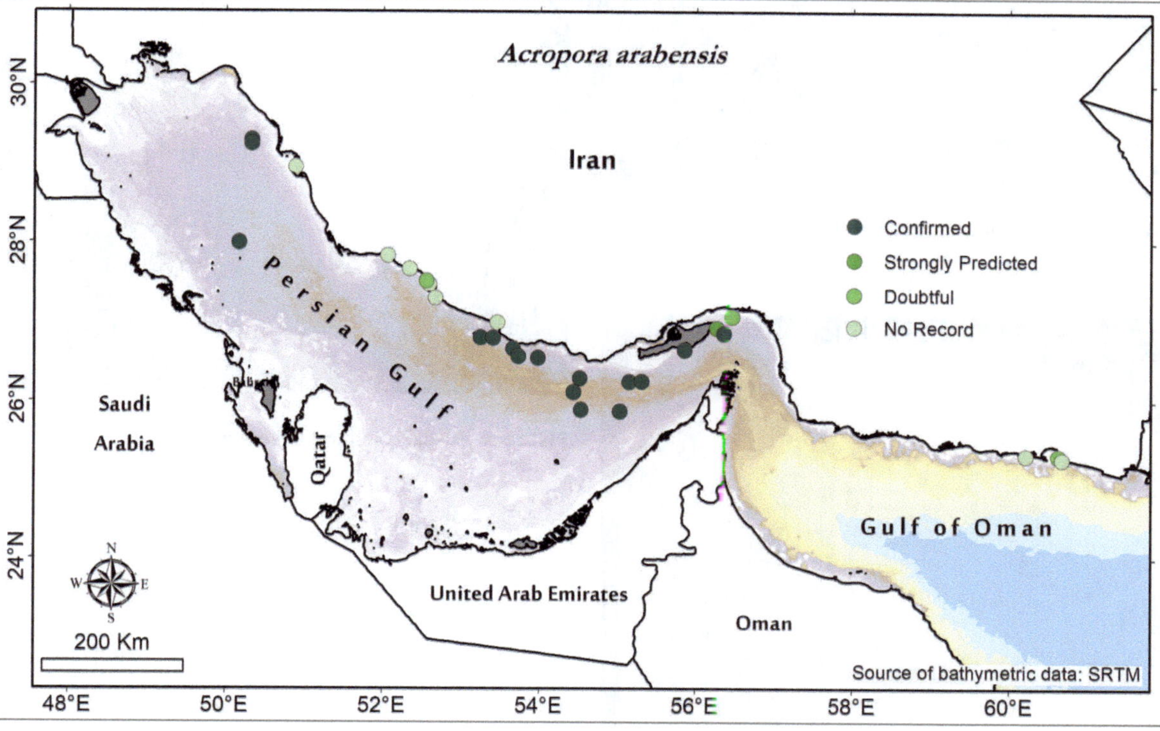

Classification: Phylum—Cnidaria, Class—Hexacorallia, Order—Scleractinia, Family—Acroporidae

Colony form: Colonies are 30 to 80 cm in diameter and have digitate to caespitose and sometimes low arborescent growth form (Plate 3.2). Branches are upright with 10 to 20 cm long and often dividing infrequently with an untidy bushy appearance. Branches are 0.5 to 1 cm in diameter near the base and taper towards the tip.

Axial Corallites: Conspicuous and dome-shaped, usually exsert and conical with outer diameter of 5–7 mm and inner diameter of about 1 mm (Plate 3.3). With six to seven primary septa developed up to 1/4 R, and 6–7 secondary septa usually present and up to 1/3 R.

Radial Corallites: Strongly appressed with small openings and thick walls (Plate 3.4). Radial corallites are tubular, with oblique or rounded apertures near the tips of the branches and nariform to dimidiate at the base of the branches. The primary septa up to 1/4 R, and the secondary septa are usually absent. Columella is not developed.

Coenosteum: On radials is sometimes reticulate with simple spines and arranged in lines. Coenostome between radials are reticulate or spinulose.

Colour: Usually reddish brown.

Distribution and abundance: This species co-occurs with other *Acropora* species and is uncommon to moderately common throughout the entire Persian Gulf mostly recorded from Kuwait, UAE, and Iran. In Iran is found mostly between 3 and 8 m in major coral reefs around the offshore islands across the northern Persian Gulf.

Plate 3.2 *Acropora arabensis* from Kish Island

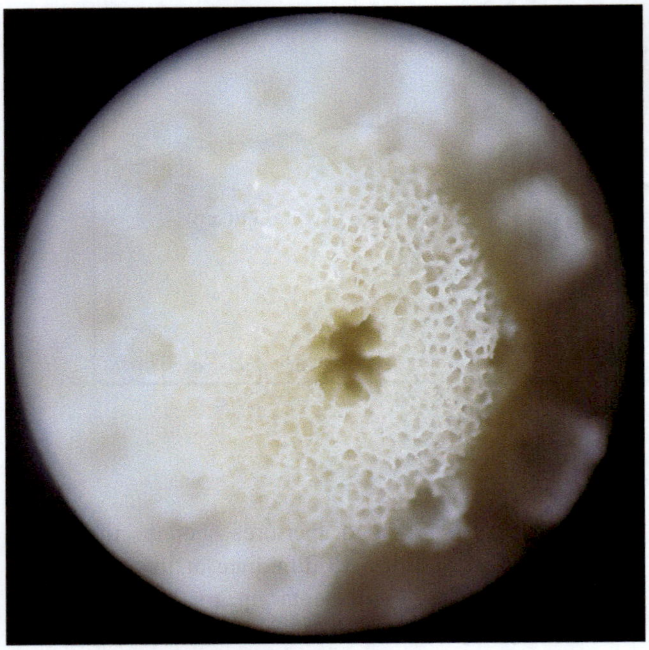

Plate 3.3 Axial corallite in *Acropora arabensis*

Plate 3.4 Radial corallites in *Acropora arabensis*

Acropora downingi (Wallace 1999)

Downing's staghorn coral			
IUCN red list assessment		Status in Iran's reefs	
Status	Population trend	Abundance	Population trend
EN	Decreasing		Decreasing

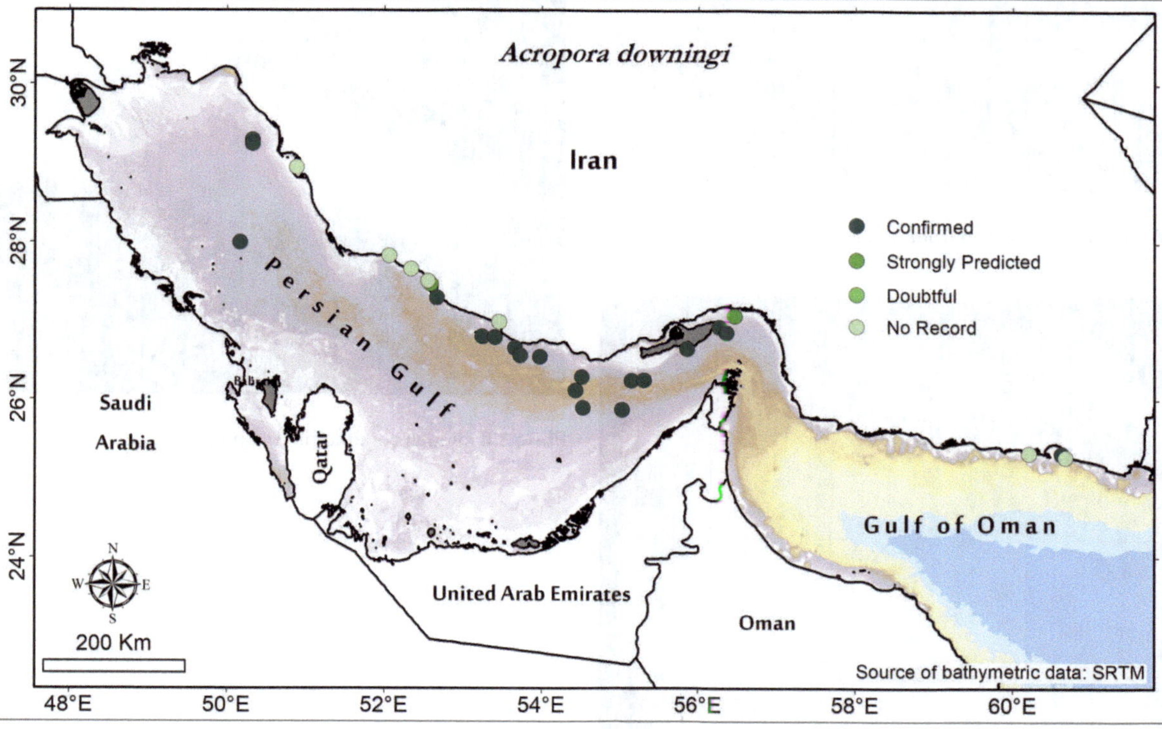

Classification: Phylum—Cnidaria, Class—Hexacorallia, Order—Scleractinia, Family—Acroporidae

Colony form: Predominantly large arborescent-tubular with a flat top, 30 to 150 cm in diameter (Plates 3.5 and 3.6). The branches mainly horizontal, from which numerous branchlets (up to 50 mm long) emerge strongly subvertical, rarely vertical. The degree of development of subvertical branchlets can vary greatly, but branchlets spacing is usually regular. The horizontal branches sometimes combine and form a plate from which branchlets with an angle between 60 and 90 degrees emerge.

Axial Corallites: Clearly developed and exsert with outer diameter of 1.5–3 mm and inner diameter of 1–1.5 mm (Plate 3.7).

Radial Corallites: Often appressed in variety of shapes and sizes and crowded with sharp-edged nariform openings (Plate 3.8). Septa in radial corallites often in two orders, with well developed (up to 2/3R) primary septa that always present, and secondary septa up to 1/4R, not always developed. Columella is not developed.

Coenosteum: On radials costate or rows of laterally compressed spines, between radials reticulate or spinulose.

Colour: Reddish and dark brown.

Distribution and abundance: This species is the most common branching coral throughout the entire Persian Gulf and Gulf of Oman and co-occurs with all other *Acropora* species. In Iran is found mostly between 3 and 8 m in major coral reefs around the offshore islands across the northern Persian Gulf and in Chabahar Bay in Gulf of Oman.

Plate 3.5 *Acropora downingi* from Abu Musa Island

Plate 3.6 *Acropora downingi* from Abu Musa Island

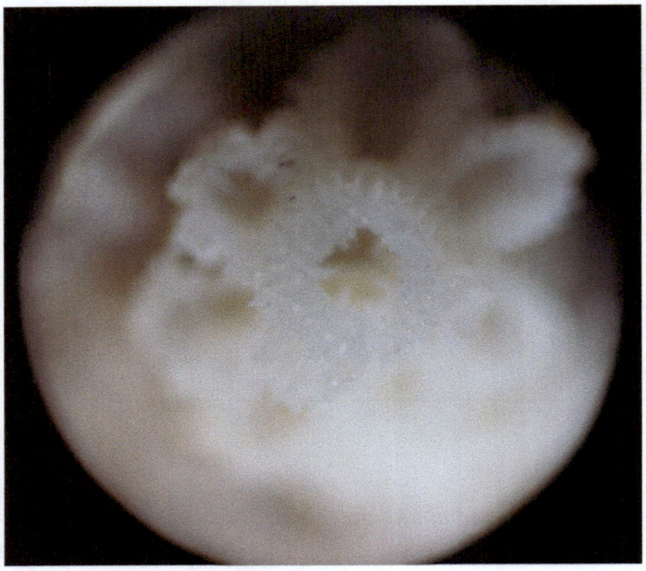

Plate 3.7 Axial corallite in *Acropora downingi*

Plate 3.8 Radial corallites in *Acropora downingi*

Acropora hyacinthus (Dana, 1846)

Brush coral			
IUCN red list assessment		Status in Iran's reefs	
Status	Population trend	Abundance	Population trend
EN	Decreasing		Decreasing

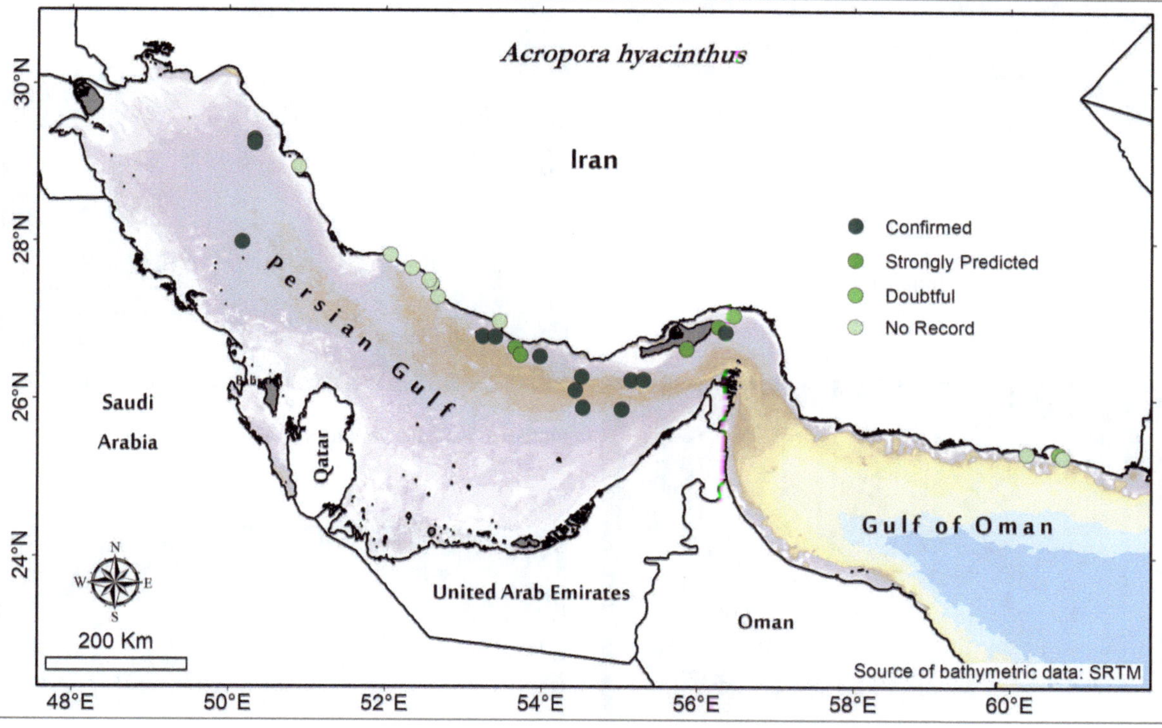

Classification: Phylum—Cnidaria, Class—Hexacorallia, Order—Scleractinia, Family—Acroporidae

Colony form: Wide plate- or table-shaped or have terraced aggregates of small plates (Plate 3.9). The diameter of the colonies ranges from 30 to 150 cm. Colonies attach to the substratum through a thick and short stem varying in height from 30 to 40 cm long, with a plate protruding from this stem. Branchlets are short, thin and relatively uniform in size and orientated upward. They are finely structured in calm water, but become closely fused when exposed to wave action. Tertiary branching order is absent.

Axial Corallites: Distinct, not exsert, porous, with two synapticular rings, outer diameter 1.5–2 mm; inner diameter 0.8–1 mm; primary septa to ¾ R (Plate 3.10).

Radial Corallites: Small, form rosettes and are cup-like (Plates 3.11 and 3.12). Most radial corallites are large with neat rounded dimidiate/lipped openings, and there are smaller corallites in between. Inner wall in radial corallites is not developed and have two synapticular rings. Primary septa are to ½ R.

Coenosteum: Costate on radial corallites and reticulate with simple spinules in between.

Colour: Light to dark brown.

Distribution and abundance: This species co-occurs with all other *Acropora* species and is relatively common throughout the entire Persian Gulf and Gulf of Oman. In Iran is found mostly between 3 and 8 m in major coral reefs around the offshore islands across the northern Persian Gulf.

Plate 3.9 *Acropora hyacinthus* from Kish Island

Plate 3.11 Radial corallites in *Acropora hyacinthus*

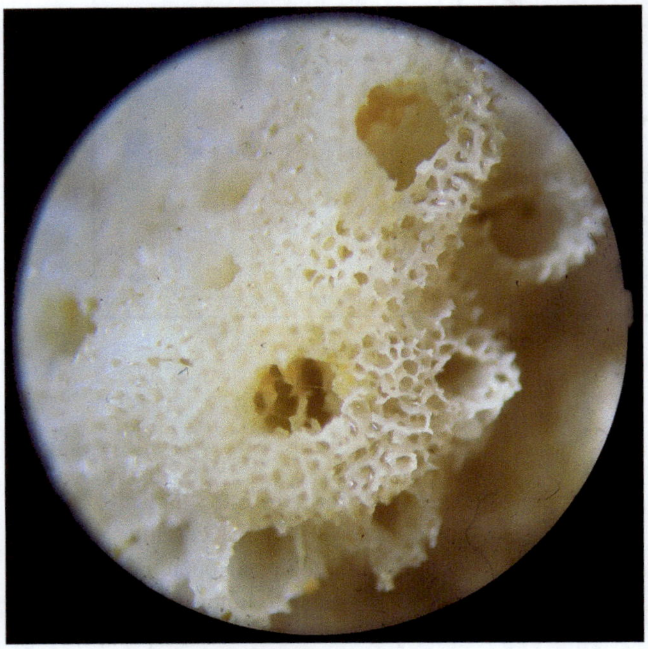

Plate 3.10 Axial corallite in *Acropora hyacinthus*

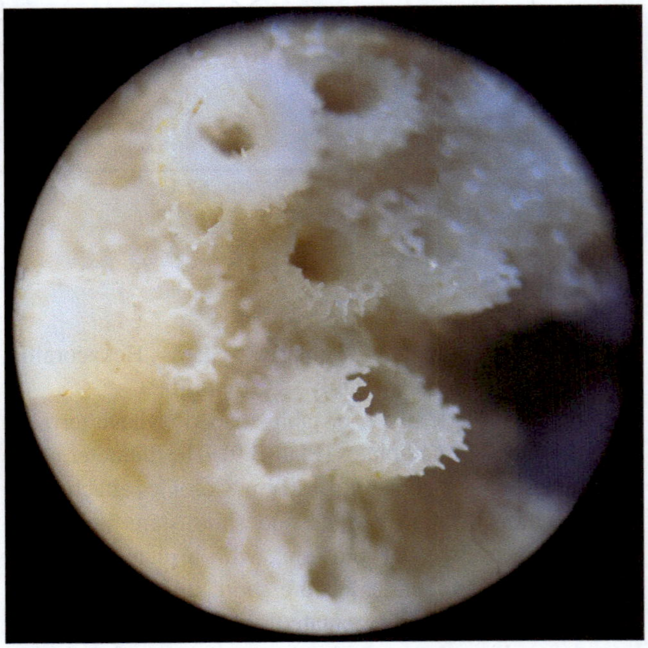

Plate 3.12 Radial corallites in *Acropora hyacinthus*

Acropora muricata (Linnaeus, 1758)

Staghorn coral			
IUCN red list assessment		Status in Iran's reefs	
Status	Population trend	Abundance	Population trend
EN	Decreasing		Decreasing

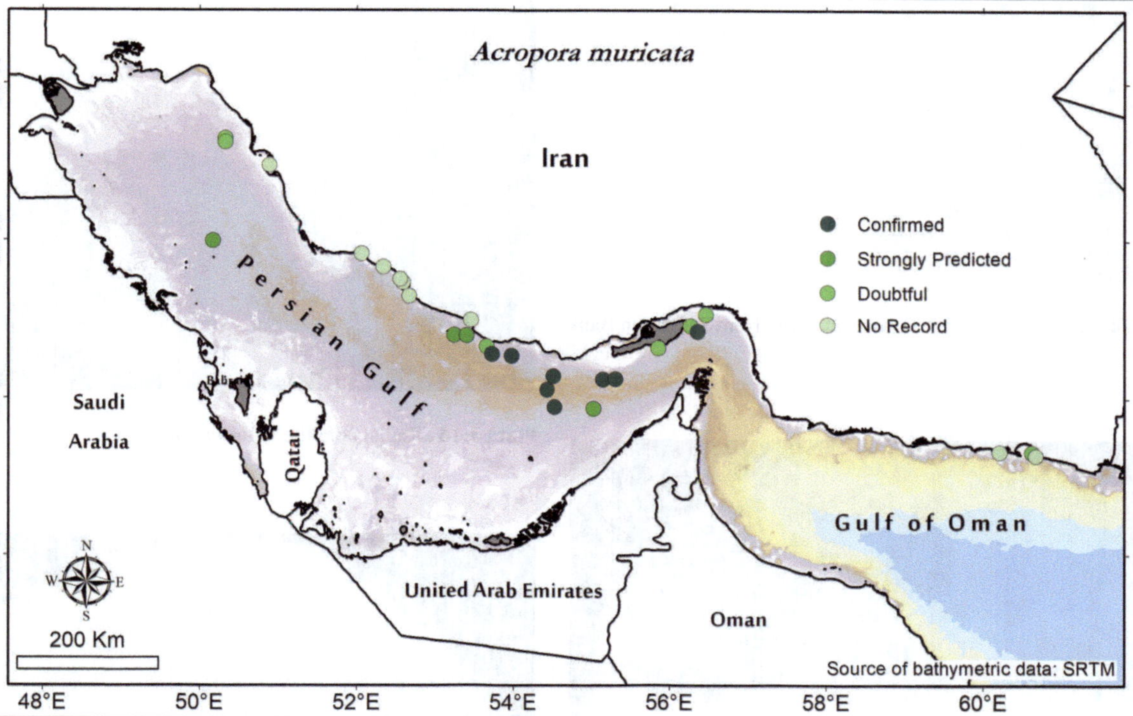

Classification: Phylum—Cnidaria, Class—Hexacorallia, Order—Scleractinia, Family—Acroporidae

Colony form: Predominantly arborescent with cylindrical branches often forming large dense aggregates or found in single formation (Plate 3.13). The diameter of the colonies ranges from 20 to 70 cm. The branches slightly tapering towards the tip and may grow to 30 cm in length with a base diameter of about 20 mm. The tertiary branching is absent.

Axial Corallites: Exsert, not porous, dominated, tapering, with outer diameter of 1.5–3.0 mm; and inner diameter of 0.8–1.2 mm (Plate 3.14). Axial corallites have two synapticular rings, and primary septa to ½ R.

Radial Corallites: Similar or different in size, crowded, tubular with small round to oblique openings, uniformly or irregularly distributed, with two synapticular rings in one size or graded (Plates 3.15, 3.16, and 3.17). The inner wall of radial corallites are developed with primary septa to ¼ R.

Coenosteum: Costate on radials; and spinule shape with single point. Reticulate in between radials.

Colour: Creamy, or light brown, generally with the pale bluish-purple branch tips.

Distribution and abundance: This species is uncommon and co-occurs with all other *Acropora* species. In Iran is found mostly between 3 and 8 m around the offshore islands towards the Strait of Hormuz in the northern Persian Gulf.

Plate 3.13 *Acropora muricata* from Sambarun Bank

Plate 3.14 Axial corallite in *Acropora muricata* from Sambarun Bank

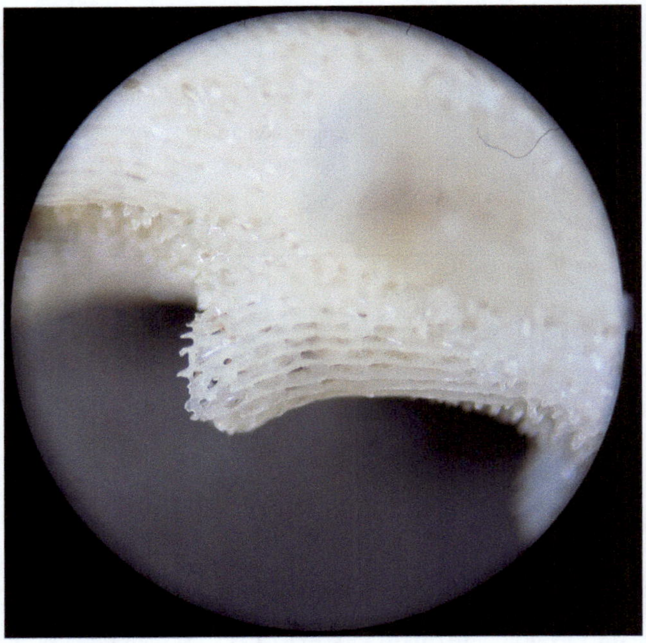

Plate 3.16 Radial corallite in *Acropora muricata*

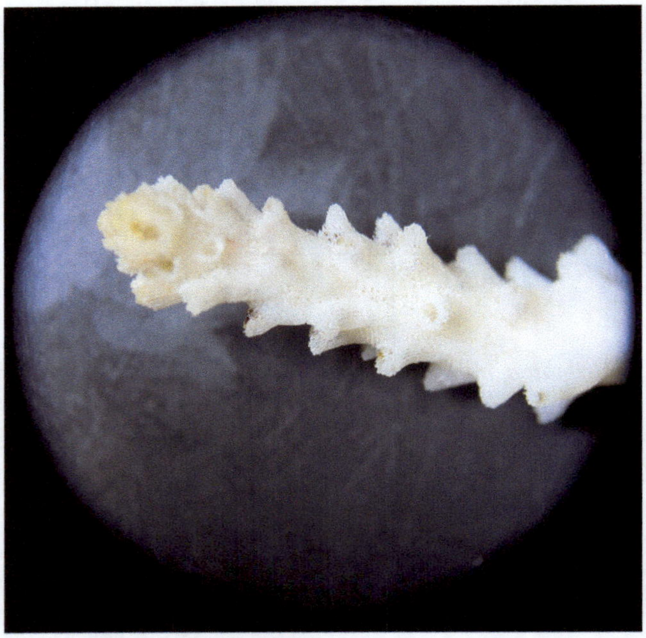

Plate 3.15 Radial corallites in *Acropora muricata*

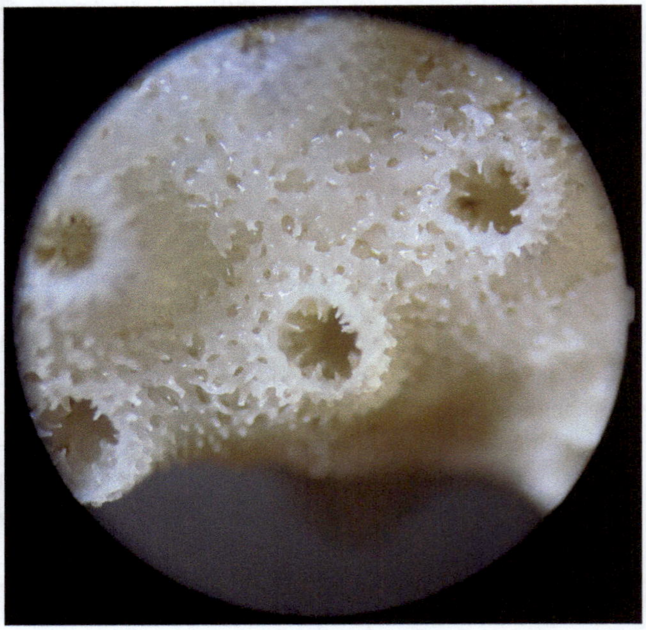

Plate 3.17 Radial corallites in *Acropora muricata*

Acropora microphthalma (Verrill, 1869)

Small-Polyped staghorn coral			
IUCN red list assessment		Status in Iran's reefs	
Status	Population trend	Abundance	Population trend
EN	Decreasing		Decreasing

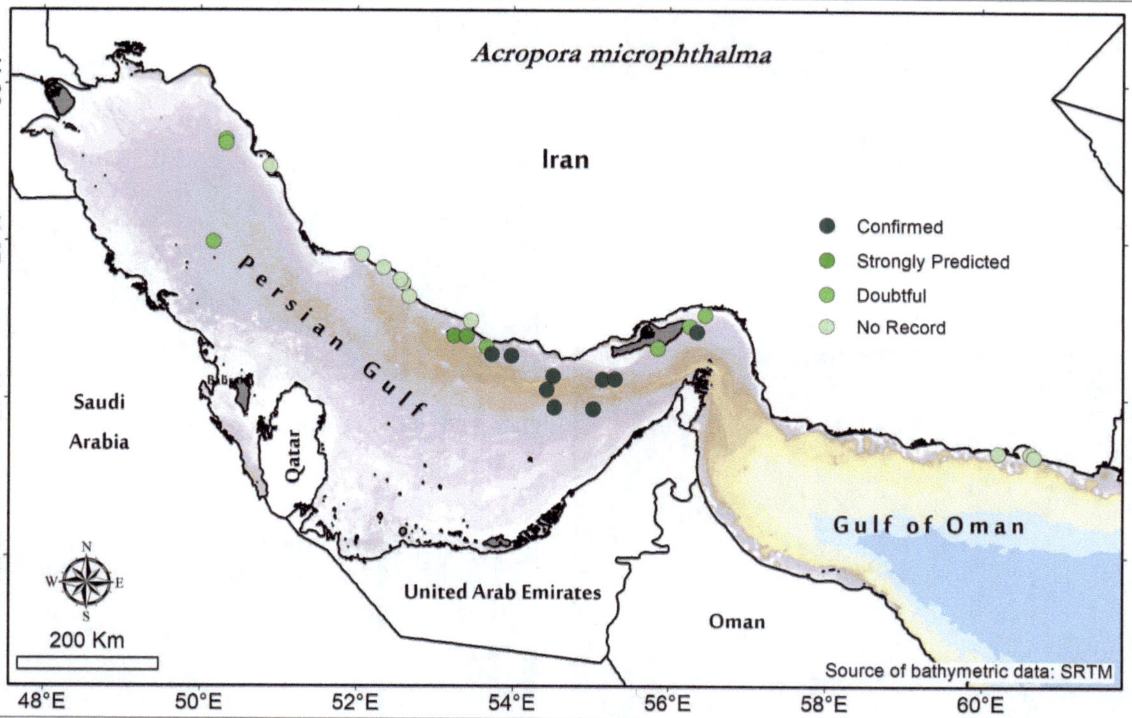

Classification: Phylum—Cnidaria, Class—Hexacorallia, Order—Scleractinia, Family—Acroporidae

Colony form: Predominantly arborescent with cylindrical, straight, tapering and frequently divided branches often forming compact thickets (Plate 3.18). The tertiary branching is present. Branches may be open with spaced branchlets or compact with sub-branches formed at acute angles. The diameter of the colonies ranges from 20 to 50 cm.

Axial Corallites: Short and tubular with two synapticular rings (Plate 3.19). with well-developed 6–7 primary septa to ¾ R, while the secondary septa may be present or incomplete. Outer diameter in axial corallites is 2.3–2.8 mm; and inner diameter 0.9–1.0 mm (Measure again).

Radial Corallites: Small, numerous, usually evenly sized and aligned in rows (Plates 3.20 and 3.21). With tubular or nariform openings, and oval-rounded towards branch tip. Inner wall developed with primary septa to 2/3 R.

Coenosteum: Identical on and between radials, costate, spinule shape, irregular and blunt.

Colour: Pale grey or sometimes pale brown or creamy.

Distribution and abundance: This species is uncommon. In Iran is found mostly in depths more than 8 m around the offshore islands and reef banks towards the Strait of Hormuz in the northern Persian Gulf.

Plate 3.18 *Acropora microphthalma* from Sambarun Bank

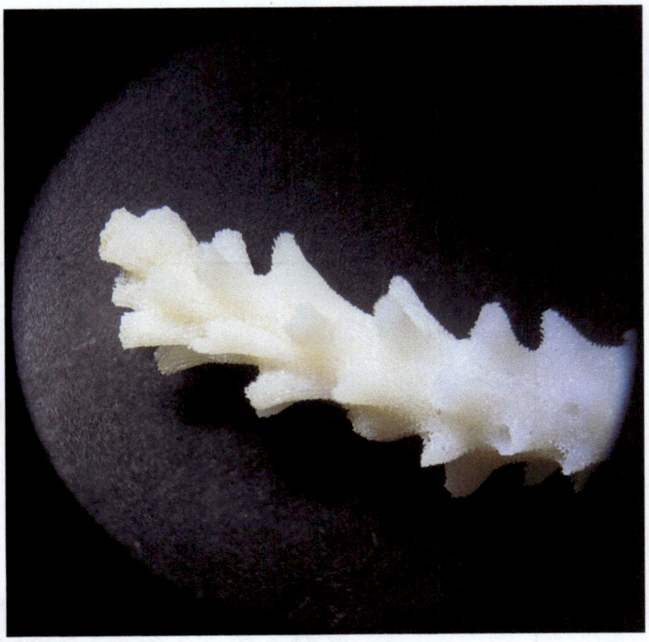

Plate 3.20 Radial corallites in *Acropora microphthalma*

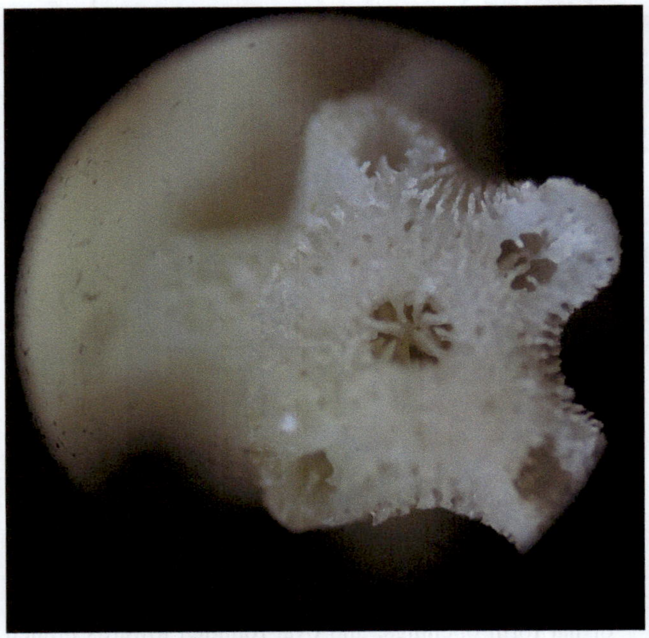

Plate 3.19 Axial corallite in *Acropora microphthalma*

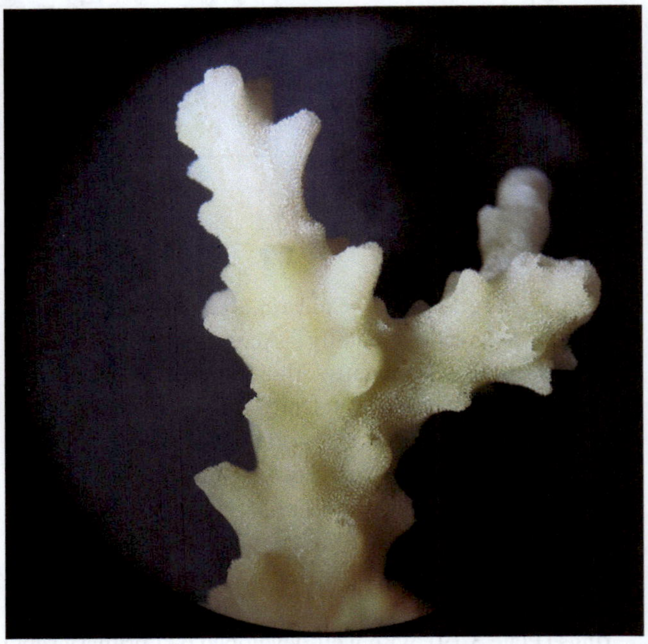

Plate 3.21 Radial corallites in *Acropora microphthalma*

Acropora valida (Dana, 1846)

Bush coral			
IUCN red list assessment		Status in Iran's reefs	
Status	Population trend	Abundance	Population trend
EN	Decreasing		Decreasing

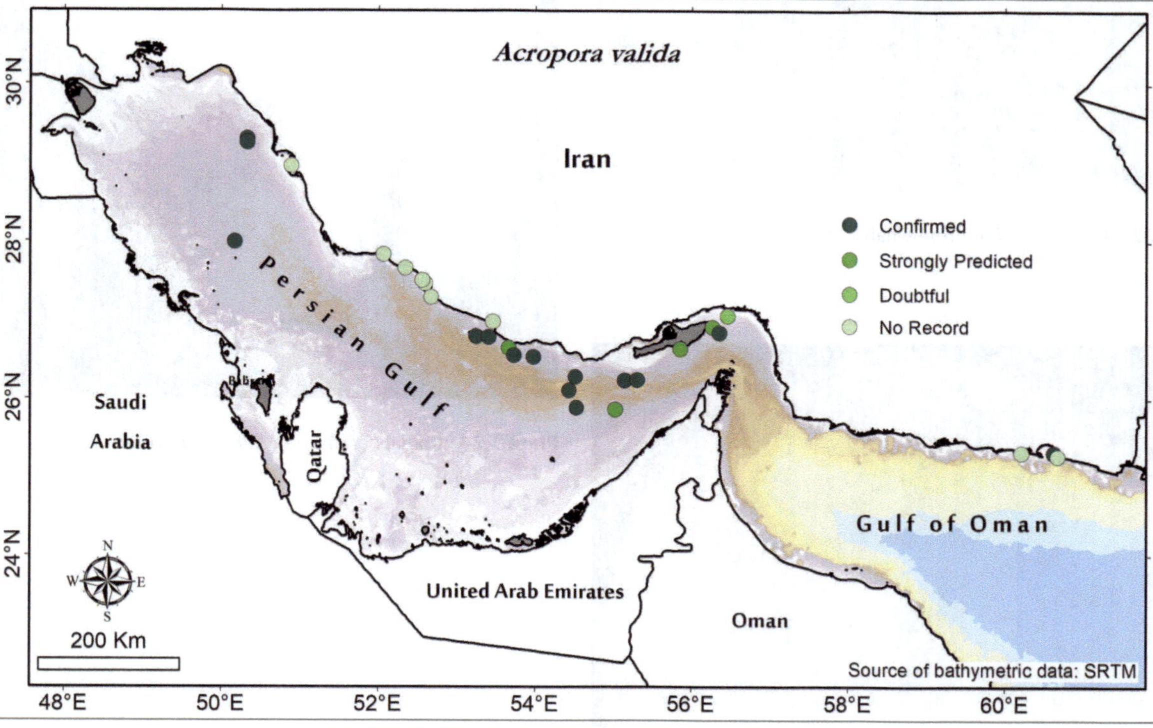

Classification: Phylum—Cnidaria, Class—Hexacorallia, Order—Scleractinia, Family—Acroporidae

Colony form: Variable growth forms from corymbose to caespitose (Plate 3.22). Colonies are small rarely growing larger than 0.5 m in diameter and may found isolated or in moderate single-species stands.

Axial Corallites: Not porous with small round openings (Plate 3.23). Outer diameter 1.6–2.8 mm and inner diameter 0.5–0.9 mm. With three synapticular cycles. Primary septa to 2/3R; secondary septa 1/3R; and tertiary septa rarely present.

Radial Corallites: Medium, appressed tubular with oval rounded openings and 2.0–3.3 mm in length (Plate 3.24). With three synapticular cycles, one size or graded with developed inner wall. Primary septa 2/3R and secondary septa 1/3R.

Coenosteum: Same on and between radials, reticulate, with densely arranged lines of spinules, and laterally flattened.

Colour: Pale brown or yellow, with cream corallites.

Distribution and abundance: This species is relatively uncommon. In Iran it is found mostly in shallow depths around the offshore islands across the northern Persian Gulf and in Chabahar Bay in Gulf of Oman.

Plate 3.22 *Acropora valida* from Chabahar Bay

Plate 3.24 Radial corallites in *Acropora valida*

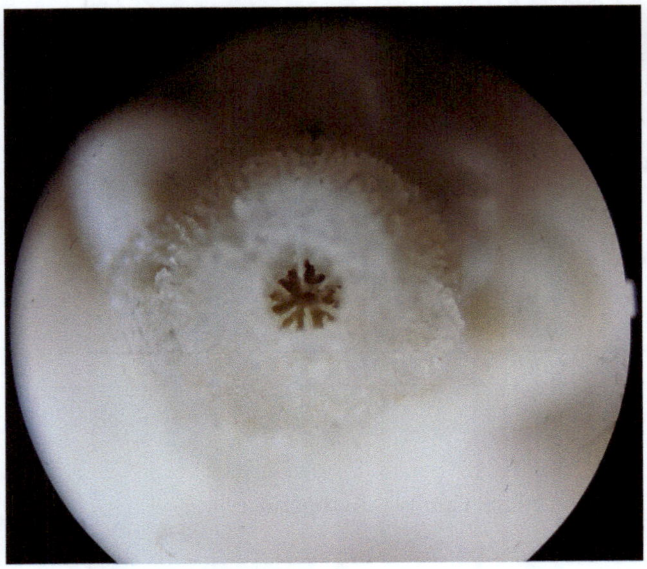

Plate 3.23 Axial corallite in *Acropora valida*

Acropora sp.

IUCN red list assessment		Status in Iran's reefs	
Status	Population trend	Abundance	Population trend
–	–		Unknown

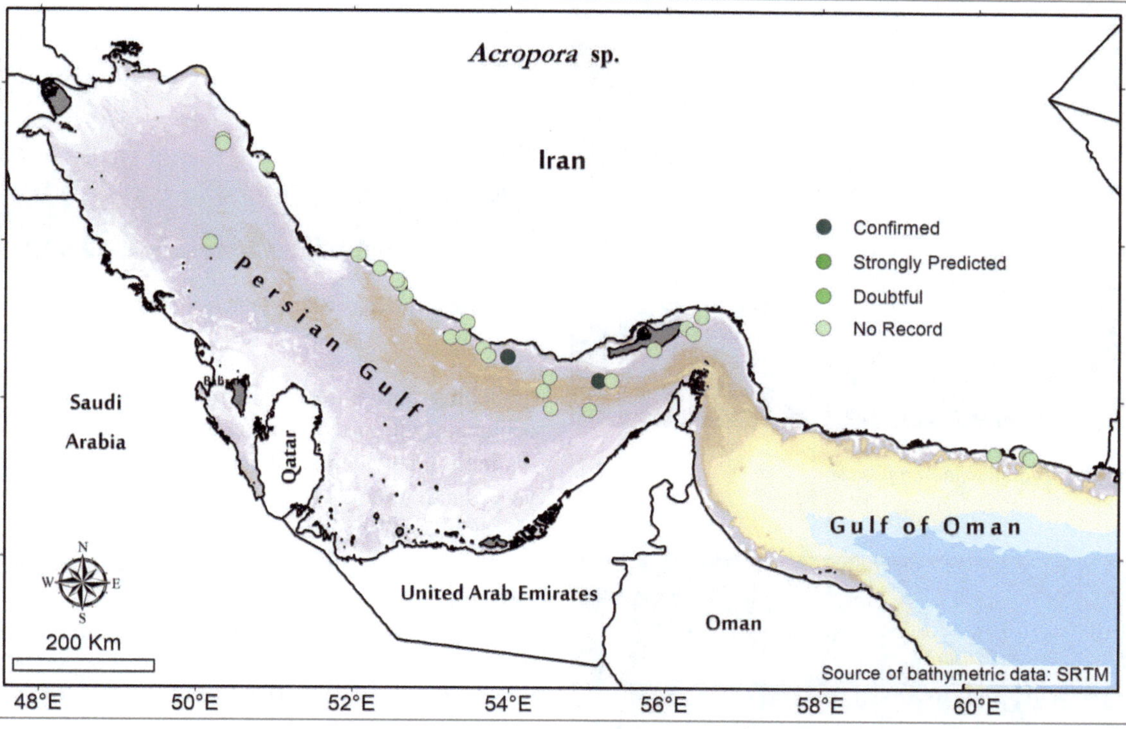

Classification: Phylum—Cnidaria, Class—Hexacorallia, Order—Scleractinia, Family—Acroporidae

Colony form: The colony exhibits a distinctly irregular growth pattern with a prominent encrusting base (Plate 3.25). Central portions are dominated by thick, conical, and upright branches, while the peripheral areas display thinner, prostrate branches that often curve upwards at their tips. A marked morphological contrast exists between the central and peripheral branches within a single colony, suggesting adaptive plasticity in branch formation depending on position and likely light or flow exposure.

Axial Corallites: Axial corallites are short, tubular, and relatively inconspicuous, yet are distinguishable from radial corallites by their consistently rounded apertures (Plate 3.26). Internally, they exhibit two complete cycles of dentate septa with two prominent, blade-like directive septa. Measurements indicate outer diameters range from 2.1 to 3.7 mm and inner diameters from 0.75 to 1.8 mm.

Radial Corallites: Radial corallites show considerable variation in size and form across the colony (Plates 3.27 and 3.28). On the primary branches, they are generally immersed to sub-immersed with a rough, rasp-like appearance, and typically possess only a single septal cycle. On the terminal branchlets, however, corallites become more prominent—tubular, rounded, and slightly appressed—with clearly dimidiate openings. Some radial corallites on extended branchlets exhibit a distinctive hook-shaped upper lip. These may display two septal cycles, though the second is often incomplete, and typically one or two directive septa are present.

Coenosteum: Neatly costate on corallites, and reticulate with many simple spinules in between. The skeletal structure is highly calcified, giving the branches a notably hard and robust character, which contributes to the colony's mechanical strength and resistance to breakage.

Plate 3.25 *Acropora* sp. from Kish Island

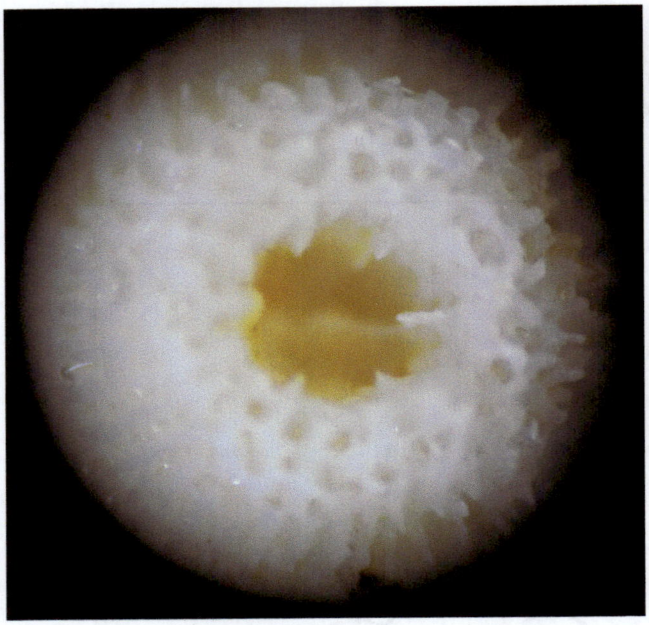

Plate 3.26 Axial corallite in *Acropora* sp.

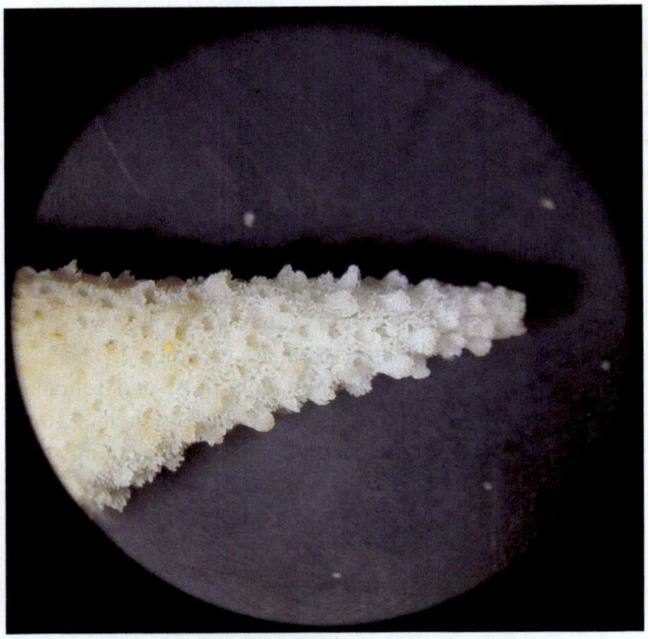

Plate 3.27 Radial corallites in *Acropora* sp.

Colour: Light brown or yellow-brown.

Distribution and abundance: This species is uncommon and very rare. In Iranian reefs it is found mostly in shallow depths around the offshore islands across the northern Persian Gulf.

3.3.1.2 Genus *Montipora* de Blainville 1830

With 72 species (Veron 2000), *Montipora* is the second largest genus in Acroporidae family. The colonies in *Montipora* genus is characterized by a wide range of corallum morphologies ranging from submassive to laminar, encrusting,

Plate 3.28 Radial corallites in *Acropora* sp.

branching and foliose, often with different growth forms occurring within the same colony. In this genus there is no distinction between axial and radial corallites. Corallites are very small (about 1 mm in diameter) and despite being expanded, polyps are hardly visible during the day in the field. Polyps are often expanded during the night. Corallites have two cycles of septa, in which the first order septa are reduced to inward-projecting spines. Columellae are absent, and the coenosteum are porous and elaborate and may form tubercles or verrucae. The species in *Montipora* genus are often confused with those in *Porites* genus. The variability in coenosteum features and colony growth form are common tools in the identification of species in *Montipora*.

Given the limited distinctive skeletal features and morphological variations, there is much confusion about the exact number of *Montipora* species in the Persian Gulf and Gulf of Oman. Thirteen *Montipora* species (i.e., *M. verrucosa*, *M. venosa*, *M. mollis*, *M.* efflorescens, *M. tuberculosa*, *M. monasteriata*, *M. foliosa*, *M. stellata*, *M. aequituberculata*, *M. circumvallate*, *M. informis*, *M. spongiosa*, *M. cf. danai*) have been reported from the entire Persian Gulf, Gulf of Oman, and Arabian Sea (Coles 1996; Claereboudt 2006; Riegl et al. 2012). Yet, records of *Montipora* have decreased over the past two decades, with no recent records from Qatar (SCENR (Supreme Council for Environment and Natural Reserves of Qatar) 2007; Sheppard et al. 2010a) and UAE (Riegl et al. 2012). Across the northern Persian Gulf, a total of six *Montipora* species have been recorded including one species from Kuwait (i.e., *M. aequituberculata*) (Hodgson and Carpenter 1995) and six species (i.e., *M. danae*, *M. spongiosa*, *M. spumosa*, *M. aequituberculata*, *M. incrassate*, *M. tuberculosa*) from the Iranian islands closer to the Straits of Hormuz (Fatemi and Shokri 2001; Shojae et al. 2010; Vajed Samiei et al. 2013; Moradi et al. 2014; Salimi et al. 2018). One *Montipora* species also is recorded from the reefs in the Chabahar Bay in the Gulf of Oman (Aminrad and Azini 2013). In the present study only one species (i.e., *M. aequituberculata*) was recorded.

Montipora aequituberculata **(Bernard, 1897)**

Encrusting pore coral			
IUCN red list assessment		Status in Iran's reefs	
Status	Population trend	Abundance	Population trend
EN	Decreasing		Decreasing

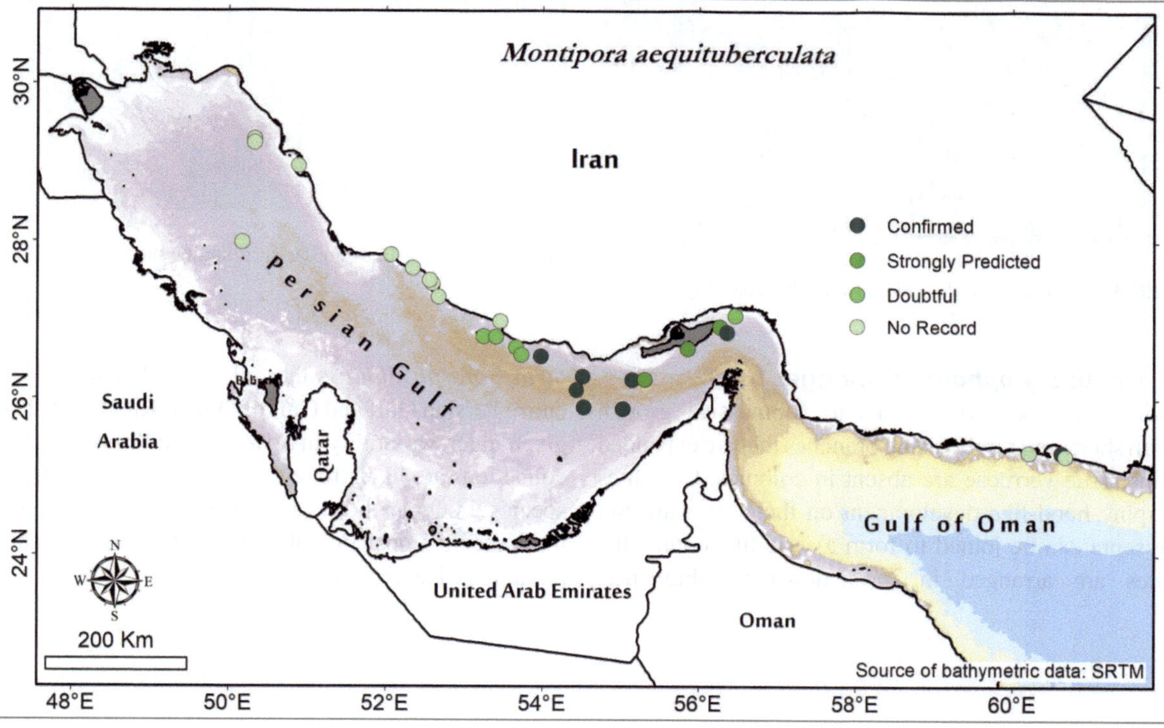

Classification: Phylum—Cnidaria, Class—Hexacorallia, Order—Scleractinia, Family—Acroporidae

Colony form: Colonies up to 80 cm in diameter, encrusting to foliose growth form, with tiers of overlapping whorls (Plate 3.29).

Corallites: The corallites are small, immersed or exsert, less than 0.5 mm, recessed within elongated thecal papillae (Plate 3.30).

Coenosteum: Reticulate, well developed and covered in elongate reticulum papillae giving corallum surface a grainy appearance.

Colour: Uniformly brown, cream, or reddish brown, pale or white on corallum margins.

Distribution and abundance: This species is uncommon and very rare. In Iran it is found mostly in shallow depths around the offshore islands in the Persian Gulf closer to the Straits of Hormuz. It is also recorded in Chabahar Bay in the Gulf of Oman.

3.3.2 Family Pocilloporidae Gray 1842

The family Pocilloporidae are colonial and worldwide are represented by 30 species in three genera including Stylophora, Pocillopora, and Seriatopora. Colonies are sub-massive, ramose or arborescent and are covered in verrucae or tubercles. The branches are irregular and often flattened or fine. Corallites are tiny and immersed and have poorly developed septa and columella. The skeleton varies widely depending on the position of corallites on branches. The coenosteum is coated with spinules. This family appears to be related to the Astrocoeniidae and Acroporidae. Representative of two genera of *Stylophora* and *Pocillopora* are recorded in the Persian Gulf and the Gulf of Oman, but *Seriatopora* is only reported from Arabian Sea (Riegl et al. 2012). The *Stylophora* is distinguished from *Pocillopora* by having a considerably larger columella and less developed septa.

Plate 3.29 *Montipora aequituberculata* from Chabahar Bay

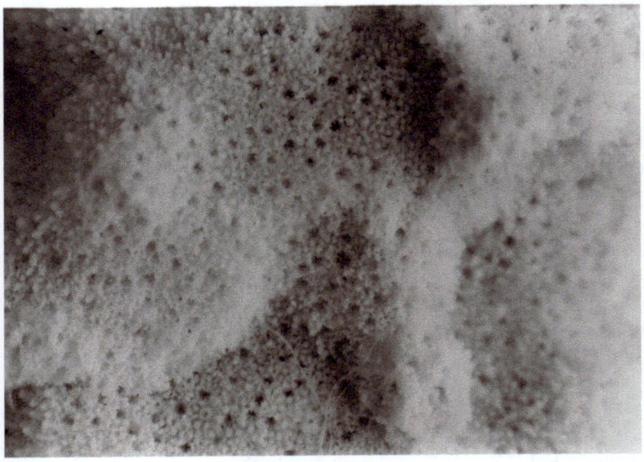

Plate 3.30 Pillar-like papillae surrounding the very small corallites

3.3.2.1 Genus *Stylophora* Schweigger 1819

Stylophora colonies have ramous to submassive growth form, with short, and rarely fusing branches that are expanded at the tips. True verrucae are absent in colonies. Corallites have a spiny hood-like development on their top wall. Six primary septa can be joined to form a style-like columella. Corrallites are arranged in rows down the branches. *Stylophora pistillata* is the only recorded species from the entire Persian Gulf and Gulf of Oman (Riegl et al. 2012). Yet, given much variability in *Stylophora* in Iranian reefs, further collections are likely to reveal the presence of more than one species. In Oman's reefs *Stylophora* has been represented by three species consisting of *S. pistillata*, S. *danae*, and *S. kuehlmanni* (Claereboudt 2006).

Stylophora pistillata (Esper, 1792)

Smooth cauliflower coral			
IUCN red list assessment		Status in Iran's reefs	
Status	Population trend	Abundance	Population trend
NT	Decreasing		Decreasing

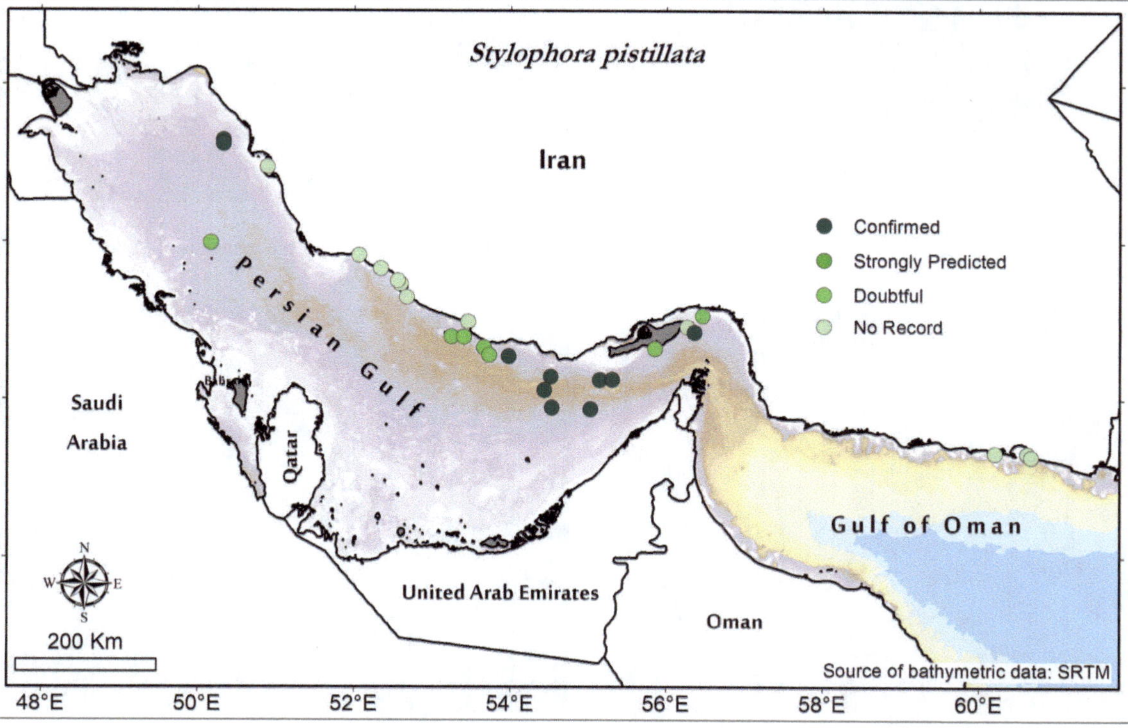

Classification: Phylum—Cnidaria, Class—Hexacorallia, Order—Scleractinia, Family—Pocilloporidae

Colony form: Colonies branching, digitate to caespitose, small mostly up to 20 cm in diameter (Plate 3.31). Branches are blunt-ended, thick and submassive, with about 10–15 mm thick (Plates 3.32 and 3.33). This species also harbors some invertebrates, including symbiotic crabs and shrimp.

Corallites: The corallites are immersed, hooded, plocoid and up to 1 mm (Plate 3.34). Columella is solid style-like (Plate 3.35). The six primary septa are short and might be fused with the columella. Sometimes there are six short secondary septa. Polyps are only extended at night.

Coenosteum: Smooth and covered with a dense arrangement of fine spinules.

Colour: Reddish brown, or greenish brown.

Distribution and abundance: This species is uncommon and very rare throughout the entire Persian Gulf. In Iran this species is mostly found in shallow depths around the offshore islands in the Persian Gulf closer to the Straits of Hormuz.

3.3.2.2 Genus *Pocillopora* Lamarck 1816

With 17 species, this genus is among the most diverse corals in the world, occurring throughout the Indo-Pacific. This is a coral genus that is common in the Arabian Sea and on Iranian islands in the Straits of Hormuz, and Chabahar Bay in Gulf of Oman. This genus is also recorded from Saudi offshore islands (Sheppard and Sheppard 1991). Yet it is very rare in the inner Persian Gulf.

Colonies are typically submassive to ramose, with branches that are often blade-like or else fine and irregular. The colonies in this genus are easily distinguished from other genera by the presence of verrucae. The corallites are immersed and typically lack clearly-defined septa and the columella is very wide, occupying most of the corallite. The corallites have a low solid columella and two unequal cycles of septa. The coenosteum is usually covered with granules. Polyps usually stretch only at night.

Genus *Pocillopora* is represented by two species (i.e., *P. damicornis*, *P. verrucosa*) in the Persian Gulf and Gulf of Oman (Coles 1996; Claereboudt 2006; Riegl et al. 2012). Likewise, in the present study both two species (i.e., *P. damicornis*, *P. verrucosa*) were recorded.

Plate 3.31 *Stylophora pistillata* from Abu Musa Island

Plate 3.33 Close-up view of the branches of *Stylophora pistillata*

Plate 3.32 Bleached colony of *Stylophora pistillata*

Plate 3.34 The conspicuous columella, six septa, and prominent hood

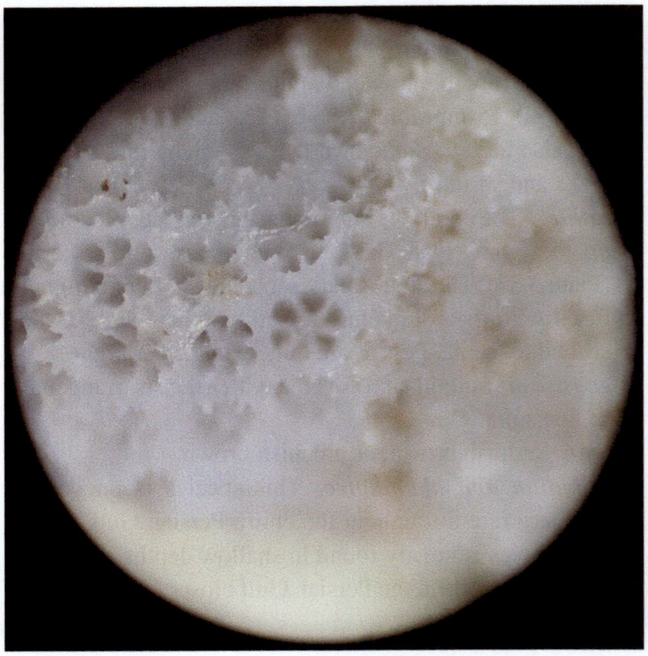

Plate 3.35 The conspicuous columella, six septa, and prominent hood

Pocillopora damicornis (Linnaeus 1758)

Cauliflower coral			
IUCN red list assessment		Status in Iran's reefs	
Status	Population trend	Abundance	Population trend
EN	Decreasing		Decreasing

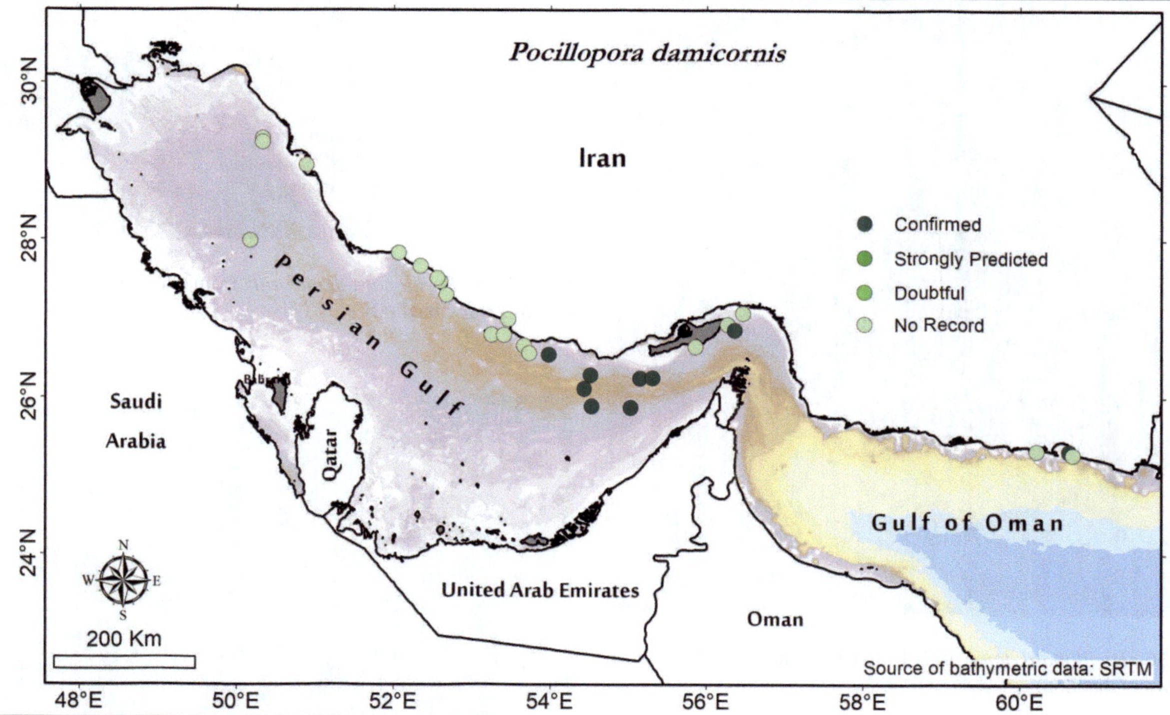

Classification: Phylum—Cnidaria, Class—Hexacorallia, Order—Scleractinia, Family—Pocilloporidae

Colony form: Colonies form compact clumps up to 30 cm high and can extend several meters across (Plates 3.36, 3.37, and 3.38). They lack true verrucae (wart-like growths), with no clear distinction between verrucae and branches as these intergrade with one another. This species is distinguished from other genus members by its irregularly arranged verrucae. It is more branched compared to the similar *P. verrucosa*. The colony's shape varies according to habitat, being thin, open, and branched in calm waters, while highly compact and sturdy in turbulent waters. The branches originate from an encrusting to submassive base.

Corallites: Branches are covered by cerioid corallites (Plates 3.39 and 3.40). The peripheral branches are much thinner than other species of this genus. Calices are small, up to 1 mm in diameter, and scattered irregularly over the colony. The distance between calices is usually less than 2 mm. Towards the tips of the branches, they become more crowded and form polygonal calices. Calices can be found all around, and are especially common on branch and branchlet tips. Septa are small and undeveloped in most corallites. The rudiments of a cycle can be found around 1/10 calice radius. Columella present, mostly highly developed and occupying most the fossa.

Colour: May be pale brown, greenish, or pink.

Distribution and abundance: This species is uncommon throughout the entire Persian Gulf. In Iran is mostly found in shallow depths in Chabahar Bay in the Gulf of Oman, and around the offshore islands in the Persian Gulf closer to the Straits of Hormuz.

Plate 3.36 *Pocillopora damicornis* from Chabahar Bay. (Photo courtesy of Amir Ghazilou)

Plate 3.37 *Pocillopora damicornis* from Chabahar Bay. (Photo courtesy of Amir Ghazilou)

Plate 3.38 *Pocillopora damicornis* from Chabahar Bay. (Photo courtesy of Amir Ghazilou)

Plate 3.39 Tip of a branch of *Pocillopora damicornis*

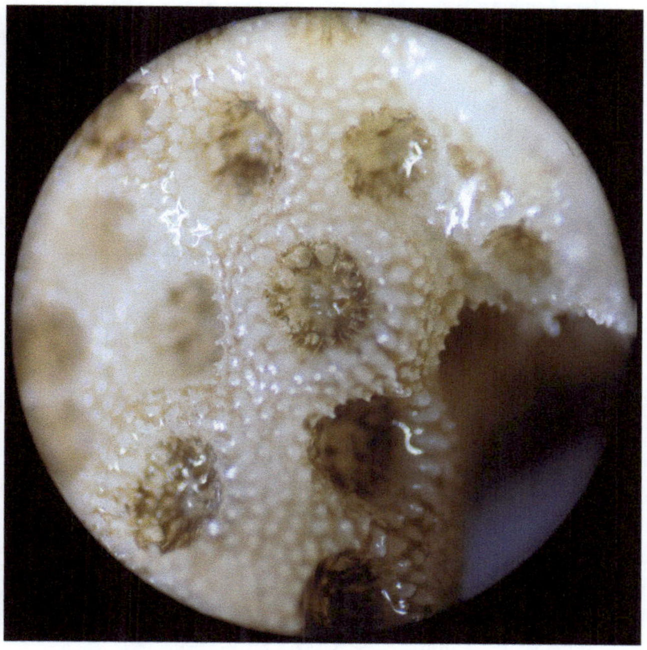

Plate 3.40 Corallites in *Pocillopora damicornis*

Pocillopora verrucosa (Ellis and Solander 1786)

Cauliflower coral			
IUCN red list assessment		Status in Iran's reefs	
Status	Population trend	Abundance	Population trend
EN	Decreasing		Decreasing

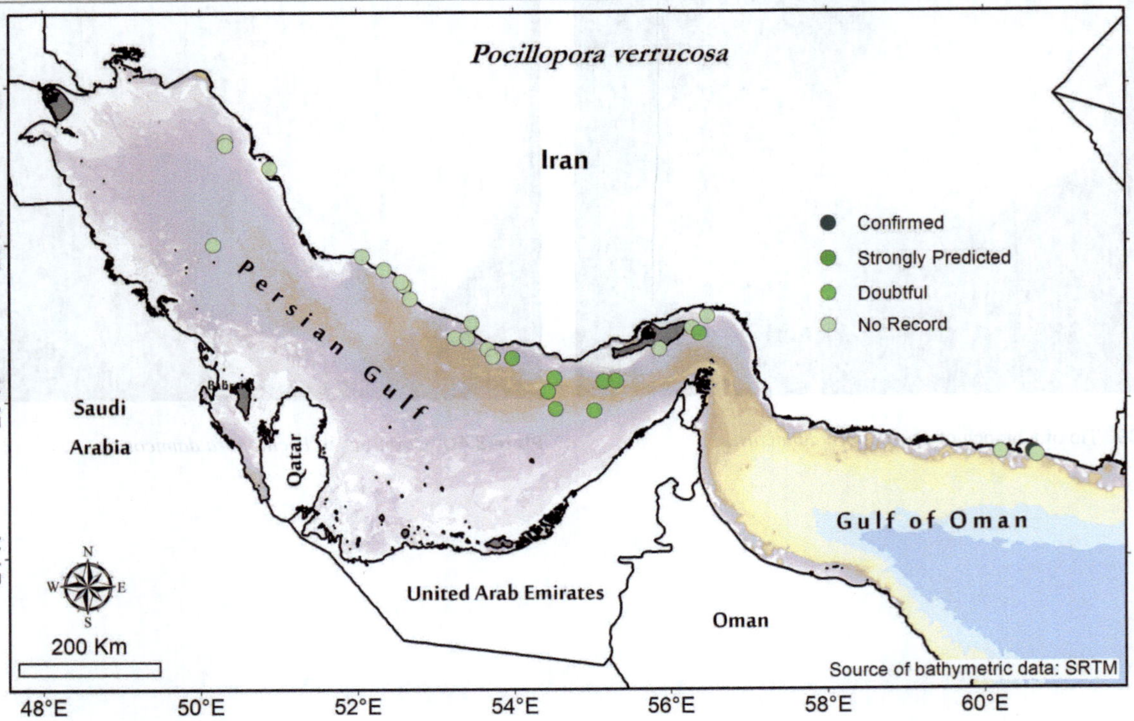

Classification: Phylum—Cnidaria, Class—Hexacorallia, Order—Scleractinia, Family—Pocilloporidae

Colony form: The colonies are composed of compact and uniform upright branches (Plate 3.41). Verrucae are prominent, irregular and large in size, giving the colony a rough surface and permanent red color.

Corallites: The branches are prominent, large, and irregular in size (Plates 3.42 and 3.43). In exposed habitats with strong waves, branches are thick and compact, while in sheltered habitats, they are open and thin. Bleached skeletons feature permanent red-brown primary branches. Calices range from 0.5 mm to 1.3 mm in diameter. Septa are irregularly sized and spaced, with two cycles. Verrucae range from 3 to 7 mm in diameter and have a height of 2 to 6 mm. The minimum branch diameter typically does not exceed approximately twice the diameter of the smallest branch. Colonies exhibit greater branching and are

Plate 3.41 *Pocillopora verrucosa* from Chabahar Bay. (Photo courtesy of Amir Ghazilou)

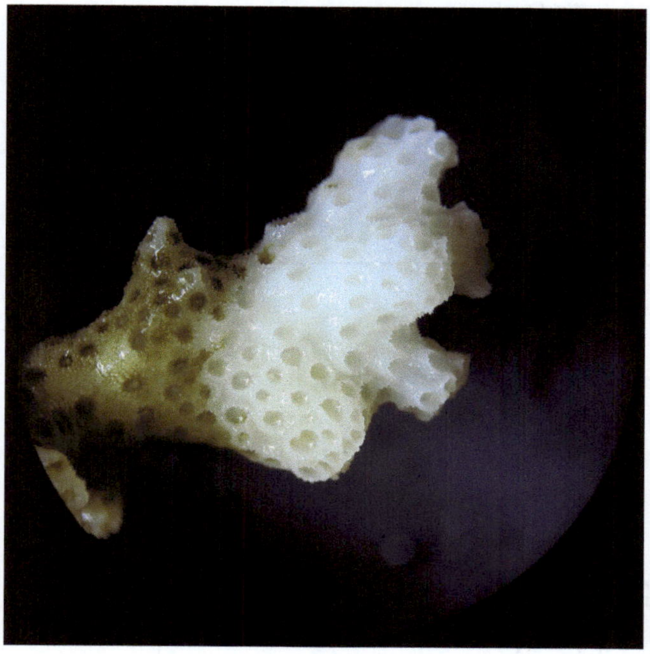

Plate 3.42 Tip of a branch of *Pocillopora verrucosa*

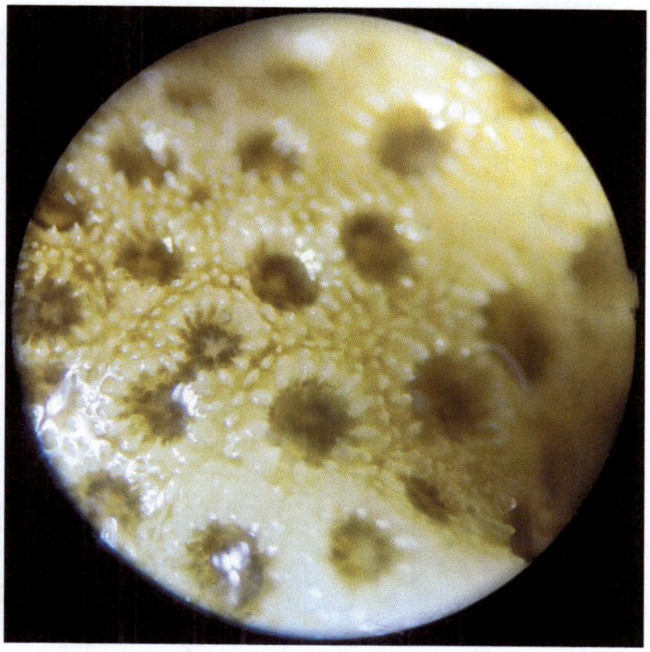

Plate 3.43 Corallites in *Pocillopora* verrucosa

less compact compared to *P. damicornis*. The branches are more delicate and less robust.

Colour: Often lighter in color, ranging from pale brown to greenish.

Distribution and abundance: This species is uncommon and very rare throughout the entire Persian Gulf. In Iran this species is only recorded from Chabahar Bay in the Gulf of Oman.

3.3.3 Family Agariciidae Gray, 1847

The family Agariciidae are mostly colonial hermatypic corals with massive, laminar or foliaceous growth forms. Corallites in this family are small and immersed with poorly defined walls formed by thickened septo-costae. Septa rarely fuse and are continuous between adjacent corallite centres. They have smooth or finely serrated margins and are closely packed. This family is represented by 47 species in six extant genera including *Agaricia*, *Dactylotrochus*, *Gardineroseris*, *Helioseris*, *Leptoseris*, and *Pavona*. *Pavona* is the only genus reported from the Persian Gulf and the Gulf of Oman (Riegl et al. 2012).

3.3.3.1 Genus *Pavona* Lamarck, 1801

Colonies have a laminar or foliaceous structure, and in some areas, they appear submassive. The corallites within the colonies are characterized by their poorly defined walls and prominently developed septocostae. These septocostae create a complex flower-like arrangement in their passage among corallites. The corallites are interconnected through protruding septo-costae. Typically, *Pavona* species tend to extend their polyps exclusively during night.

Of 14 species recorded worldwide, genus *Pavona* is represented by eight species (i.e., *P. cactus*, *P. frondifera*, *P. varians*, *P. explanulata*, *P. diffluens*, *P. venosa*, *P. decussate*, *P. duerdeni*) in the Persian Gulf and Gulf of Oman (Coles 1996; Claereboudt 2006; Riegl et al. 2012). However, three species (i.e., *P. decussate*, *P. explanulata*, *P. varians*) were recorded from the Iranian reefs across the Persian Gulf and the Gulf of Oman.

Pavona decussata (Dana, 1846)

Leaf coral			
IUCN red list assessment		Status in Iran's reefs	
Status	Population trend	Abundance	Population trend
LC	Decreasing		Stable

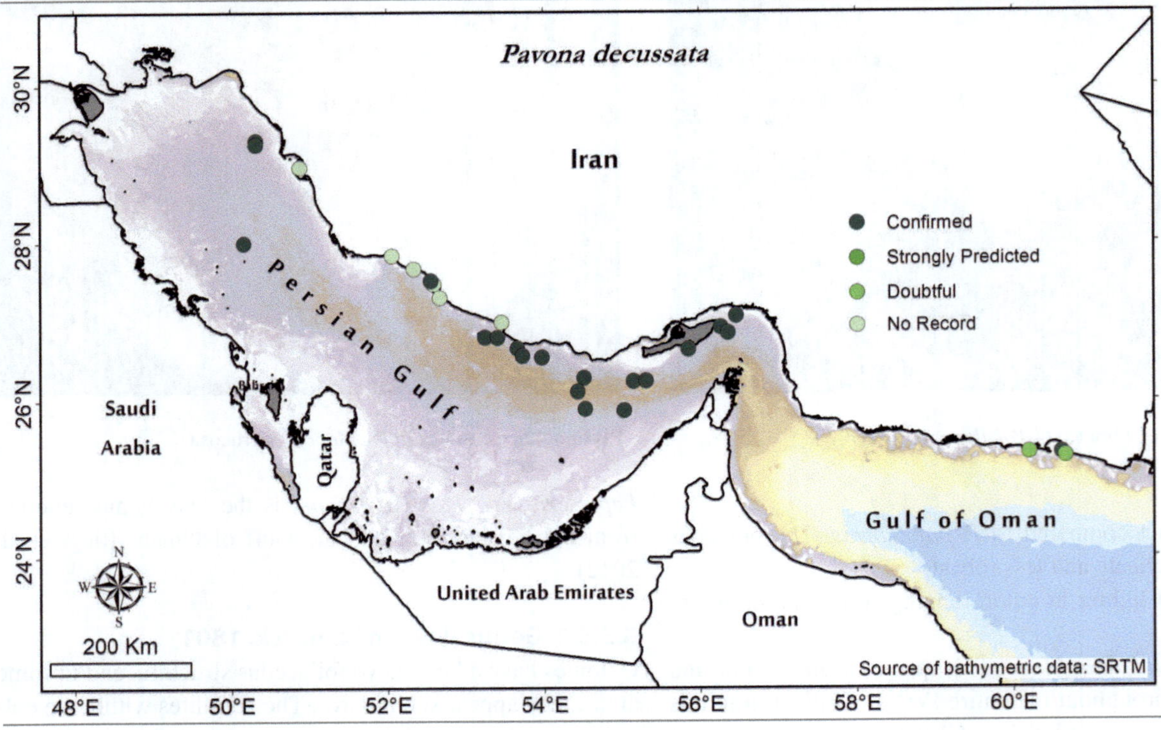

Classification: Phylum—Cnidaria, Class—Hexacorallia, Order—Scleractinia, Family—Agariciidae

Colony form: The corallum structure is characterized by interconnected and twisted, bifacial, moderately thick (0.5–1 cm) vertical plates that emerge from an encrusting, sub-massive, or even massive core (Plates 3.44 and 3.45). Notably, the fronds of the corallum lack lobed flat edges.

Corallites: The corallites of this particular species exhibit thamnasterioid morphology, characterized by diameters ranging from 2 to 3 mm (Plates 3.46 and 3.47). These corallites often display significant irregularity in size. Additionally, they exhibit cerioid growth patterns, frequently forming short to medium-long series that may include up to ten individual corallites. Notably, the corallites are generally deep, with septa descending relatively abruptly toward the central axis. Their diameters vary significantly, spanning from 3 to 7 mm, reflecting considerable variability within the species. The septa within this species exhibit a distinct arrangement, occurring in two alternating orders. The first order consistently extends to the calyx center and fuses with the columella. In contrast, the second order of septa tends to be shorter and does not consistently reach the columella; some second order septa may even split between adjacent calices. Notably, the septal sides of the first order display regular, robust granulations and serrated upper margins, while the second order septa typically appear smooth. Additionally, the septocostae are well-developed. The columella, predominantly present, often manifests as a vertical boss or plate. During daylight hours, the typically white tentacles can extend.

Colour: This species exhibits shades ranging from brown to beige.

Distribution and abundance: This species is uncommon and very rare throughout the entire Persian Gulf. In Iran this species is only recorded from Chabahar Bay in the Gulf of Oman.

Plate 3.44 *Pavona decussata* from Kharg Island

Plate 3.45 *Pavona decussata* from Abu Musa Island

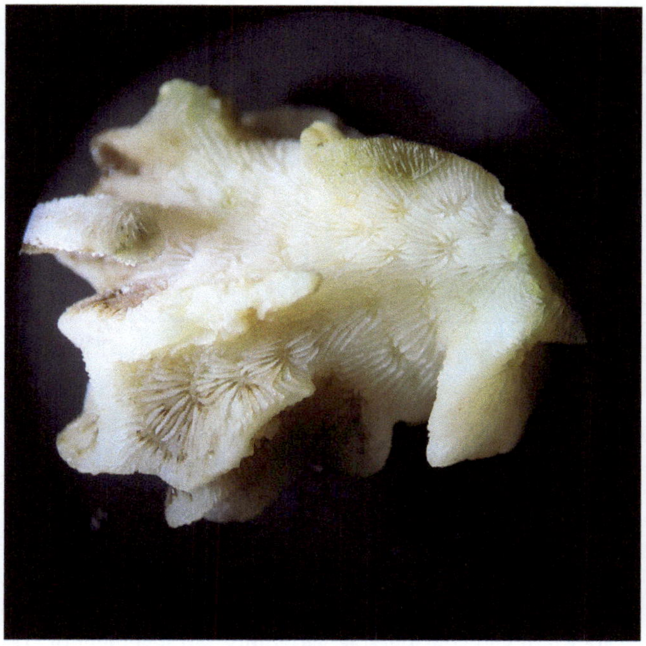

Plate 3.46 Corallites in *Pavona decussata*

Plate 3.47 Corallites in *Pavona decussata*

Pavona explanulata (Lamarck, 1816)

Plate Pavona Coral			
IUCN red list assessment		Status in Iran's reefs	
Status	Population trend	Abundance	Population trend
LC	Decreasing		Unknown

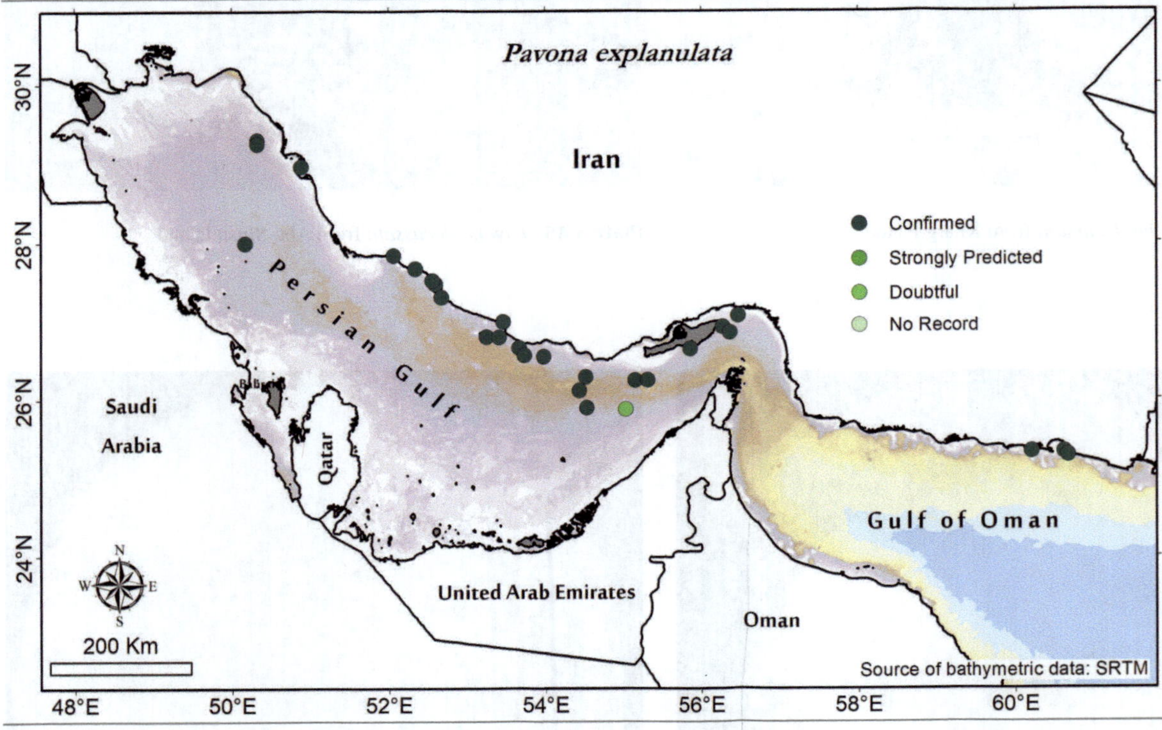

Classification: Phylum—Cnidaria, Class—Hexacorallia, Order—Scleractinia, Family—Agariciidae

Colony form: The corallum structure exhibits flat, encrusting, or laminar characteristics (Plate 3.48). In large colonies, it transitions from massive to columnar, conforming to the shape of the underlying substratum.

Corallites: Corallites are relatively small (larger than those in other Pavona species), compact, and lack any discernible corallite wall (Plates 3.49 and 3.50). They are widely spaced and circular. Notably, the coarse septa extend between calyces, alternating in thickness, resulting in a hash-marked appearance. However, this structural detail may not be readily apparent when the coral polyps are expanded. While leaves are infrequently developed, the coral typically forms an encrusting growth with minimal relief. Occasionally, delicate, leaf-like structures may occur. The calyx diameter ranges from 3 to 4 mm, with variable separation between calyces depending on water depth and clarity. Corallites exhibit a circular outline, appearing thamnasterioid, with diameters ranging from 3 to 7 mm and considerable variability. Arrangement of corallites tends to follow rows parallel to the corallum margin. The septa exhibit two alternating orders: the first order is often thicker and more exsert, always reaching the columella, while the second order is usually shorter and does not extend to the columella. Well-developed septocostae alternate within the coral. Additionally, a solid vertical boss, known as the columella, is present and well-defined.

Colour: The color of this coral species ranges from brown to beige, featuring a typical white pattern.

Distribution and abundance: This species is uncommon and very rare throughout the entire Persian Gulf. In Iran this species is only recorded from Chabahar Bay in the Gulf of Oman.

Plate 3.48 *Pavona explanulata* in Chabahar Bay

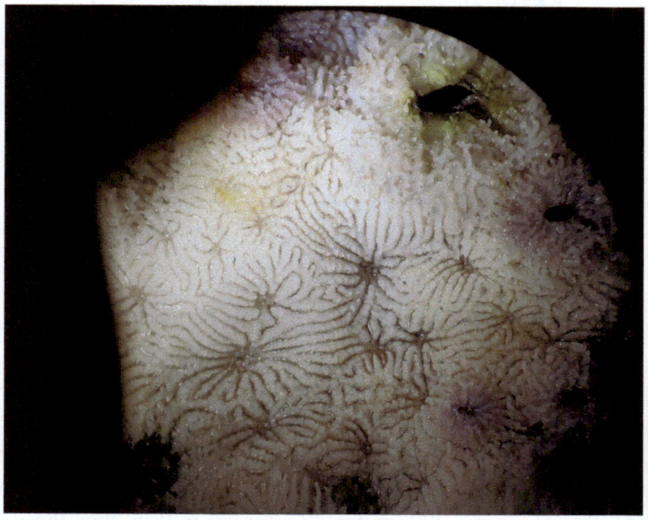

Plate 3.49 Corallites in *Pavona explanulata*

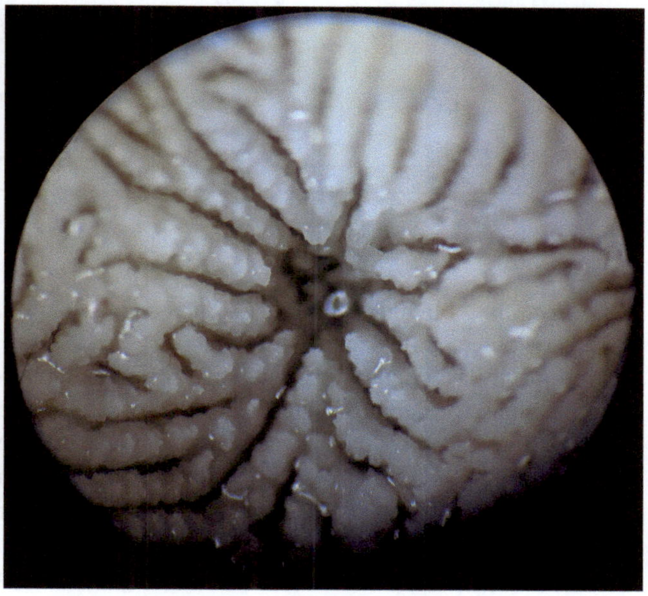

Plate 3.50 Corallite in *Pavona explanulata*

Pavona varians (Verrill, 1864)

Corrugated coral			
IUCN red list assessment		Status in Iran's reefs	
Status	Population trend	Abundance	Population trend
LC	Decreasing		Unknown

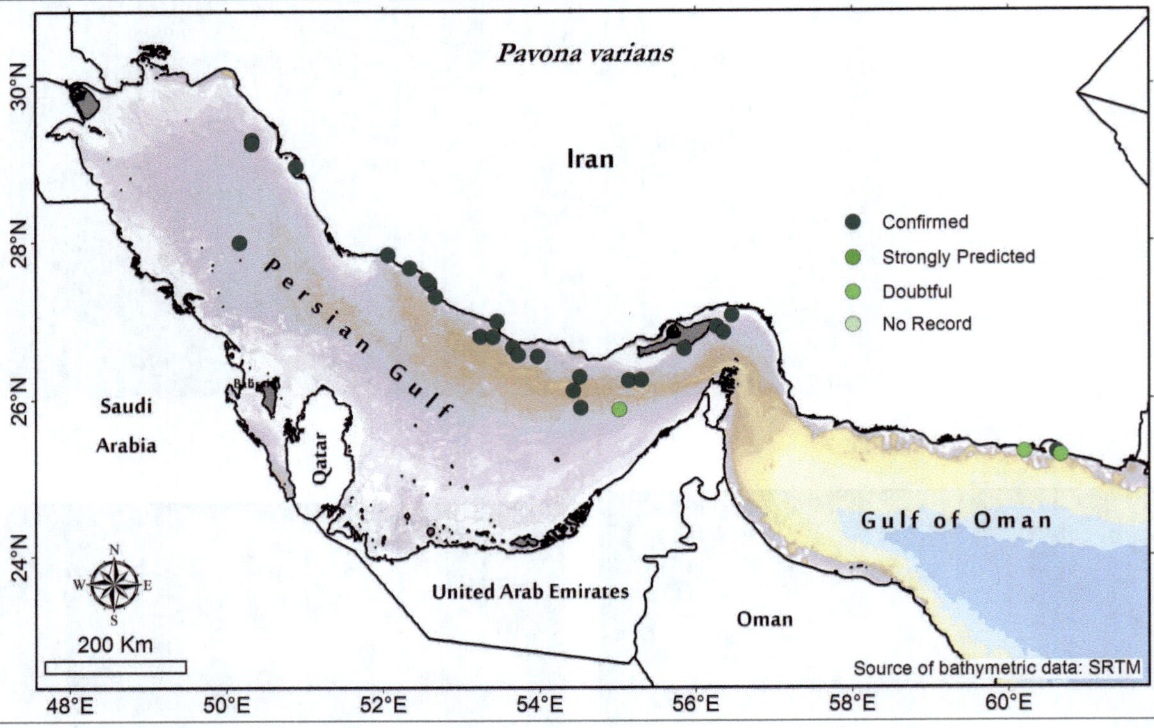

Classification: Phylum—Cnidaria, Class—Hexacorallia, Order—Scleractinia, Family—Agariciidae

Colony form: The corallum exhibits diverse forms, including submassive, laminar, or encrusting structures (Plate 3.51). These adaptive features allow it to thrive in diverse environments, where irregular valleys intersect with aligned corallites along ridges, resulting in an irregular distribution on flat surfaces.

Corallites: The corallites manifest as short, irregular valleys or align in a perpendicular arrangement between ridges (Plate 3.52). Occasionally, they even adopt a cup-like morphology. The septa within this species exhibit two alternating orders, although at times, their distinction may be subtle.

Colour: The color of this spceis ranges from pale yellow and beige to light brown or gray.

Distribution and abundance: This species is uncommon throughout the entire Persian Gulf. It is found across Iranian reefs in both the Persian Gulf and the Gulf of Oman.

3.3.4 Family Dendrophylliidae Gray, 1847

Dendrophylliidae as the third most diverse coral family primarily inhabit the Indo-Pacific region and exhibit unique characteristics. Unlike typical reef-building corals, they are often non-reef builders. Their ecological range extends to depths of up to 900 meters. Colonies within this family exhibit both solitary and colonial forms, with a predominant ahermatypic nature. The corallite walls, typically composed of coenosteum, exhibit porosity. Notably, the septa display a distinctive fusion pattern, particularly evident in immature corallites. Taxonomically distinct from other coral families. Some members within this family are azooxanthellate. This group comprises only three hermatypic genera: *Turbinaria*, *Duncanopsammia*, and *Heteropsammia*. Despite their superficial dissimilarity, these genera share a common lineage. Approximately 20 genera fall under this family's classification. Among them, four genera—*Turbinaria*, *Duncanopsammia*, *Heteropsammia*, and *Dendrophyllia* —are recorded in Iranian reefs. Globally, a total of 183 coral species have been recorded, with four species specifically identified within Iranian reefs.

Plate 3.51 *Pavona varians* from Sambarun Bank

Plate 3.52 Corallites in *Pavona varians*

3.3.4.1 Genus *Turbinaria* Oken, 1815

Colonies of *Turbinaria* exhibit a wide range of sizes, from small to large, and display distinct morphologies. In shallow waters, the colonies are predominantly foliaceous and irregularly shaped, while in deeper waters, they take on a small, mushroom-like form. The corallites within these colonies are notably large and visibly connected by an extensive coenosteum. Corallites are circular, immersed to tubular and have porous walls that mirror the surrounding coenosteum. The septa and columella are typically well-defined, with short and neat septa and broad, compact columellae. Polyps usually extended only during nighttime. *Turbinaria* is a well-defined genus, distinct from other genera except for occasional similarities with *Astrepora*. Within the Persian Gulf and Gulf of Oman, *Turbinaria* is represented by four species: *T. peltata*, *T. reniformis*, *T. mesenterina*, and *T. aurea* (Coles 1996; Claereboudt 2006; Riegl et al. 2012). Notably, only *T. reniformis* has been documented in the Iranian reefs across both regions.

Turbinaria reniformis **(Bernard, 1896)**

Yellow scroll coral			
IUCN red list assessment		Status in Iran's reefs	
Status	Population trend	Abundance	Population trend
LC	Decreasing		Decreasing

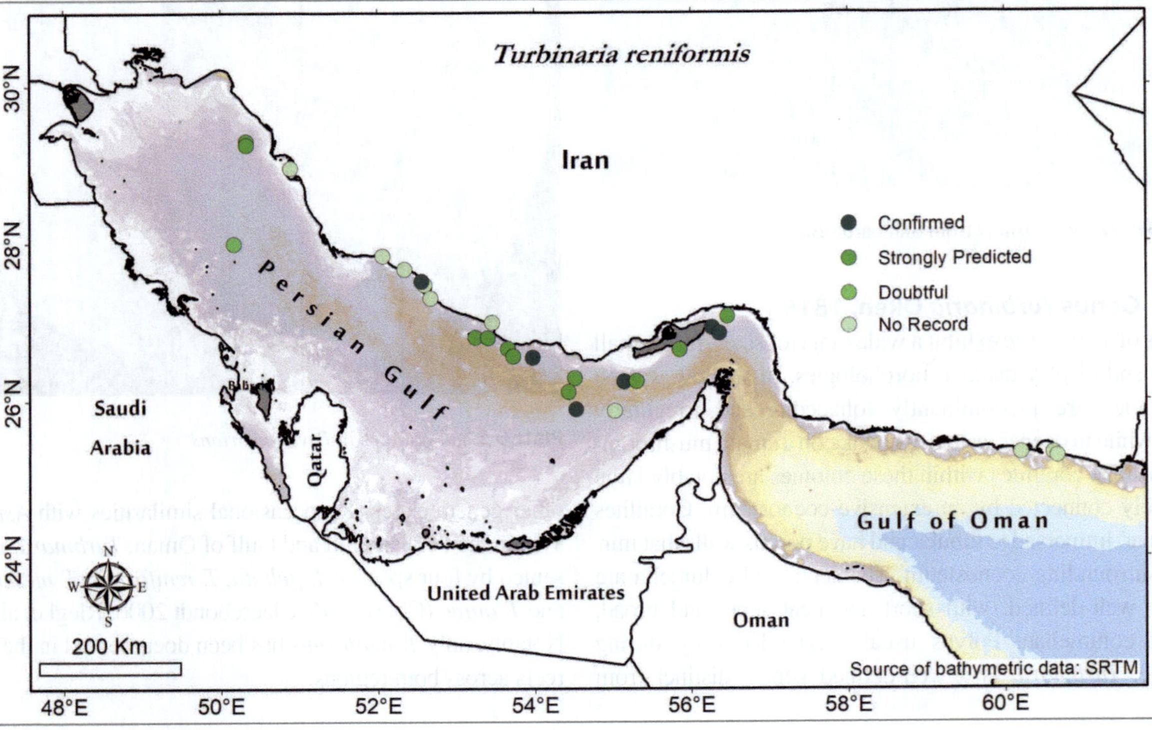

Classification: Phylum—Cnidaria, Class—Hexacorallia, Order—Scleractinia, Family—Dendrophylliidae

Colony form: The corallum of *T. reniformis* comprises large, irregularly foliate colonies that can form multi-layered sheets (Plate 3.53).

Corallites: The corallites within these colonies are more densely packed, smaller, and irregular compared to those found in *Duncanopsammia peltata* (Plates 3.54, 3.55, and 3.56). Occasionally, they are arranged in irregular rows. The septa, numbering between 12 and 20, generally follow a single order without significant differentiation. Their margins are finely dentate, and the sides exhibit pronounced granulation. The columella, although often barely visible deep within the calice, may be absent. When present it consists of vertical elements. The thecae are well-developed and spongy. It is commonly found in all shallow habitats.

Colour: Overall, the coloration and gross morphology of this species make it readily distinguishable and easy to identify. The overall color of *T. reniformis* typically ranges from light brown to orange.

Distribution and abundance: This species is frequently encountered in shallow habitats throughout the entire Persian Gulf. Within Iranian reefs is found in both the Persian Gulf and the Gulf of Oman.

3.3.4.2 Genus *Duncanopsammia* Wells, 1936

The genus *Duncanopsammia* is commonly associated with reef-building corals. Within this genus, individual polyps exhibit relatively large with round skeletal bases. These polyps display a wide, flat, circular disk, typically in light brown or green hues, often adorned with subtle stripes. *Duncanopsammia* corals thrive in tropical marine environments, particularly within the shallow habitats of the Persian Gulf and the Gulf of Oman. Their presence significantly contributes to the biodiversity of these regions. The genus comprises two species: *Duncanopsammia axifuga* and *D. peltata*. Notably, *D. peltata* is exclusively recorded in the Iranian reefs of the Persian Gulf and the Gulf of Oman.

Plate 3.53 *Turbinaria reniformis* from Sambarun Bank

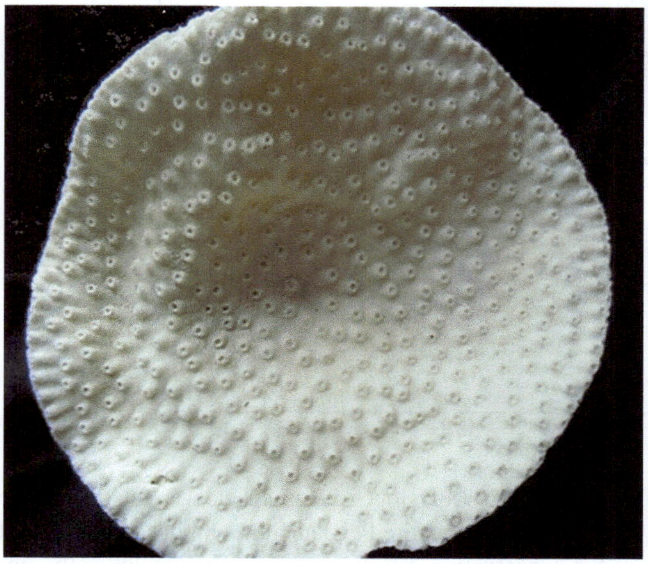

Plate 3.54 Bleached *Turbinaria reniformis*

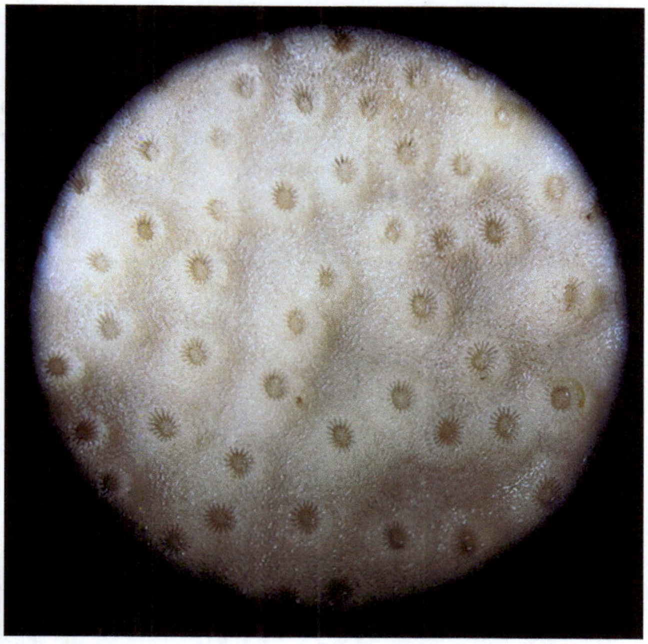

Plate 3.55 Corallites in *Turbinaria reniformis*

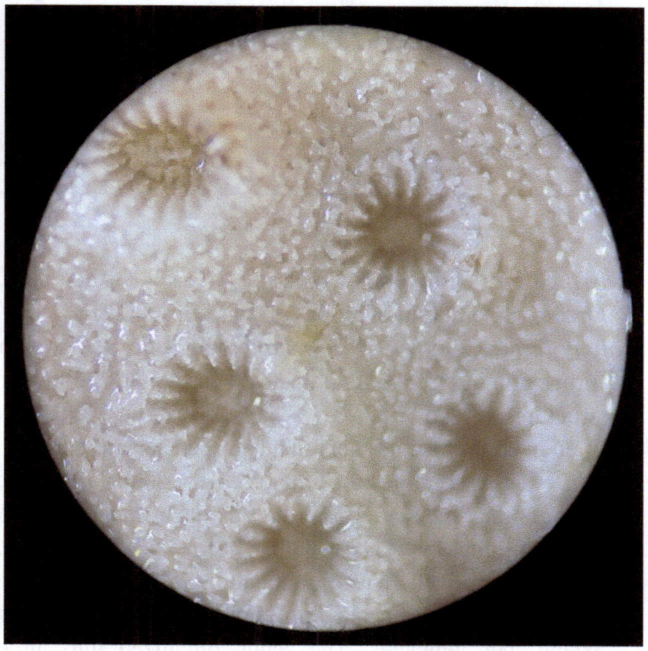

Plate 3.56 Corallites in *Turbinaria reniformis*

Duncanopsammia peltata **(Esper, 1790)**

Whisker coral			
IUCN red list assessment		Status in Iran's reefs	
Status	Population trend	Abundance	Population trend
LC	Decreasing		Decreasing

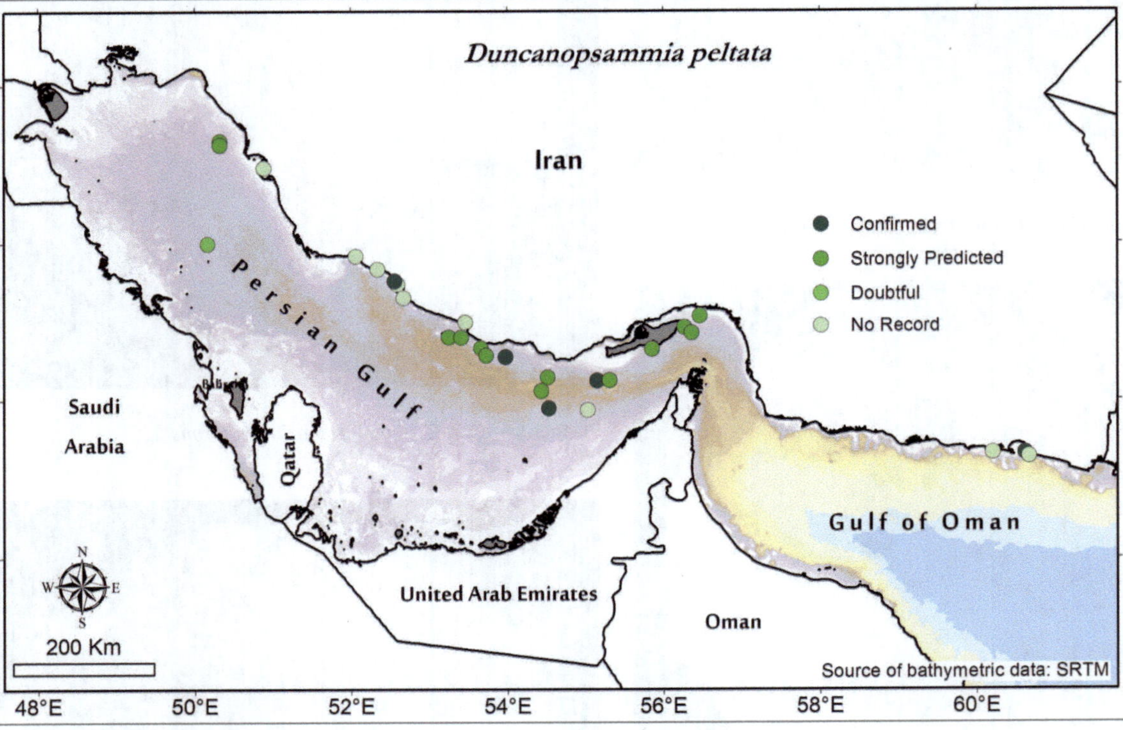

Classification: Phylum—Cnidaria, Class—Hexacorallia, Order—Scleractinia, Family—Dendrophylliidae

Colony form: Colonies of *D. peltata* exhibit diverse growth forms, ranging from flat laminae to occasionally columnar structures (Plates 3.57, 3.58, and 3.59). These colonies often overlap, forming tiers that can span several meters.

Corallites: The corallites are immersed to tubular in shape, with minimal inclination and their diameter ranges from 3 to 5 millimeters (Plates 3.60, 3.61, 3.62, and 3.63). Corallites are well spaced, maintaining a regular circular arrangement. During daylight hours, the large polyps extend their tentacles. The septa, organized into three orders, exhibit similar features in the first two orders. These septa project inwards for approximately half the radius of the corallite. The third cycle of septa is reduced and may even be absent. Septal sides display granulated textures. A well-developed columella, characterized by twisted septal dentations, contributes to the overall structure. The thecae are well developed and exhibit a spongy texture.

Colour: The typical coloration of the corallites is brown to beige, with paler polyps.

Distribution and abundance: These corals thrive in protected environments, particularly on shallow rocky foreshores with turbid water. Additionally, they can be found on shallow reef slopes. This species is relatively rare throughout the entire Persian Gulf, emphasizing the need for conservation efforts to protect this ecologically significant coral species. This species is recorded in the Iranian reefs of the Persian Gulf and the Gulf of Oman.

3.3.4.3 Genus *Heteropsammia* Milne Edwards & Haime, 1848

Heteropsammia, a hard coral well adapted to soft substrates, consistently harbors worms, suggesting a dependence on them. Globally, nine nominal species have been reported. *Heteropsammia* exhibits both solitary and colonial forms, with up to seven calices resulting from mono- or distomodaeal intratentacular budding. The base encloses a tube inhabited by commensal sipunculids. The wall of *Heteropsammia* is thick and spongy, with costae reduced to multiple rows of crispatce granulations, forming a labyrinthine pattern over the base.

The septa follow Pourtalès plan, and the lower cycles are also thick and spongy. A well-developed columella is present. Notably, *Heteropsammia* and *Heterocyathus* share both structural and ecological equivalences, while *Psammoseris* serves as at least a structural equivalent. The sole species recorded in the Iranian reefs of the Persian Gulf is *Heteropsammia cochlea*.

Plate 3.57 *Duncanopsammia peltata* in Kish Island

Plate 3.58 *Duncanopsammia peltata* from Chabahar Bay

Plate 3.59 *Duncanopsammia peltata* from Chabahar Bay

Plate 3.60 Corallites in *Duncanopsammia peltata* from Kish Island

Plate 3.61 Corallite in *Duncanopsammia peltata* from Kish Island

Plate 3.62 Corallites in *Duncanopsammia peltata* from Chabahar Bay

Plate 3.63 Corallite in *Duncanopsammia peltata* from Chabahar Bay

Heteropsammia cochlea **(Spengler, 1781)**

Walking Dendro coral			
IUCN red list assessment		Status in Iran's reefs	
Status	Population trend	Abundance	Population trend
LC	Decreasing		Unknown

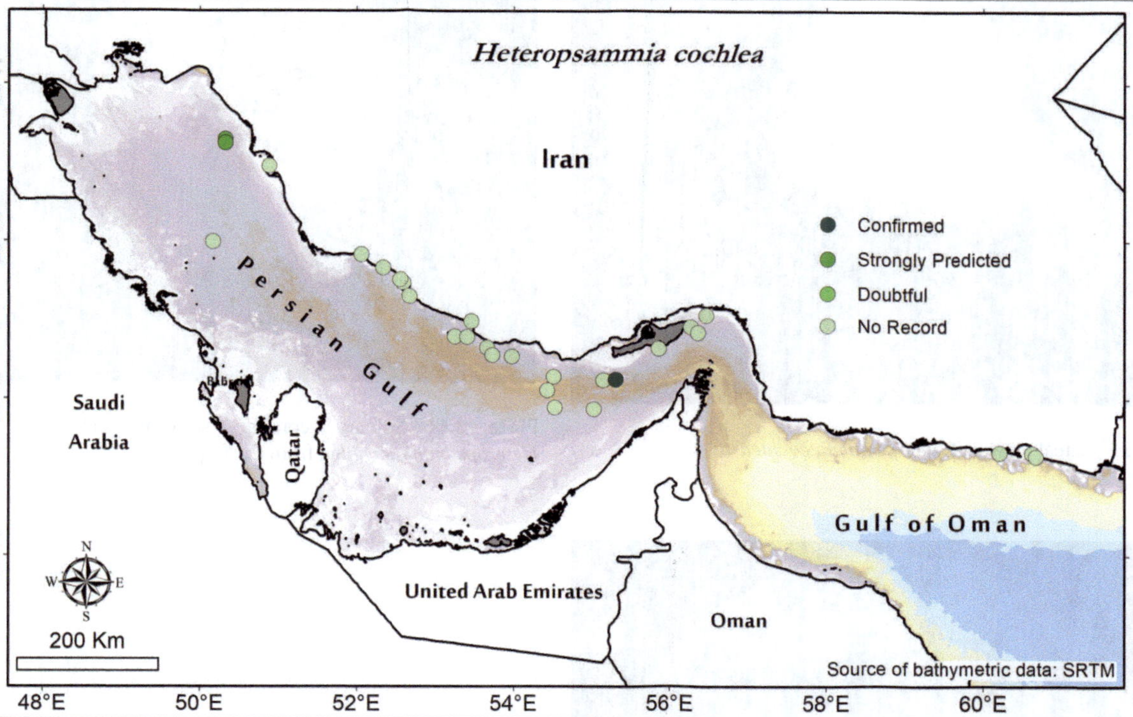

Classification: Phylum—Cnidaria, Class—Hexacorallia, Order—Scleractinia, Family—Dendrophylliidae

Colony form: This species exhibits a free-living lifestyle, their basal morphology varying between flat and keeled forms, contingent upon substrate characteristics (Plate 3.64). The corallum takes on an hourglass shape, often housing one or more corallites. They possess a central hole at the base, which serves as a refuge for a commensal sipunculid worm, aiding the coral in its movement (Plate 3.65). The sipunculid actively shifts the coral across soft substrates, preventing its burial. Individual coral colonies assume a rounded shape, with diameters of up to 25 millimeters.

Corallites: Typically, a coral colony possesses one or two corallites; in cases where only one is present, it often exhibits an hourglass shape (Plate 3.66). Certain populations form colonies with as many as 30 corallites. These corallites feature well-developed septa, adhering to the Pourtalès plan. The columellae are broad, compact, and deeply seated. The coral walls consist of a porous coenosteum. Tentacles, primarily extended during nocturnal hours, exhibit zooxanthellate associations in tropical locales, while temperate regions and deeper waters may harbor azooxanthellate forms.

Colour: The color of these corals ranges from pale grey to orange-brown or greenish hues.

Distribution and abundance: This species is not common throughout the entire Persian Gulf. In Iran this species is recorded from sandy intertidal areas in Qeshm Island in the Persian Gulf.

3.3.4.4 Genus *Tubastraea* Lesson, 1830

The majority of *Tubastraea* species are azooxanthellate, with colonies characterized by compact growth forms, which closely resemble small mounds. The individual corallites are plocoid in shape, with thin walls primarily composed of coenosteum. The septa within these corallites are thin, following the Pourtales plan, particularly during the early developmental stages. Additionally, the columellae are small in size. This genus is documented across the Persian Gulf and Gulf of Oman, with recorded occurrences of *T. coccinea* and *T. tagusensis* in Kuwait (Carpenter et al. 1997), *T. coccinea* and *T. aurea* in Iran (Riegl et al. 2012), and *T. aurea* and *T. micranth* in the Sultanate of Oman (Coles 1996; Claereboudt 2006).

Plate 3.64 Lateal view of *Heteropsammia cochlea*

Plate 3.66 Solitary corallite showing calice structures in *Heteropsammia cochlea* from Qeshm Island

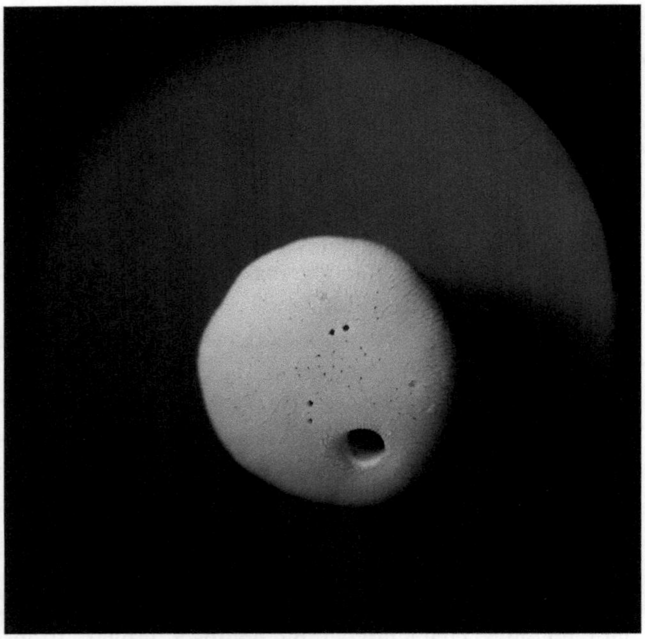

Plate 3.65 Aboral view of *Heteropsammia cochlea* showing the tiny opening from which a sipunculid worm extends its proboscis

Tubastraea coccinea (Lesson, 1830)

Orange cup coral			
IUCN red list assessment		Status in Iran's reefs	
Status	Population trend	Abundance	Population trend
–	–		Unknown

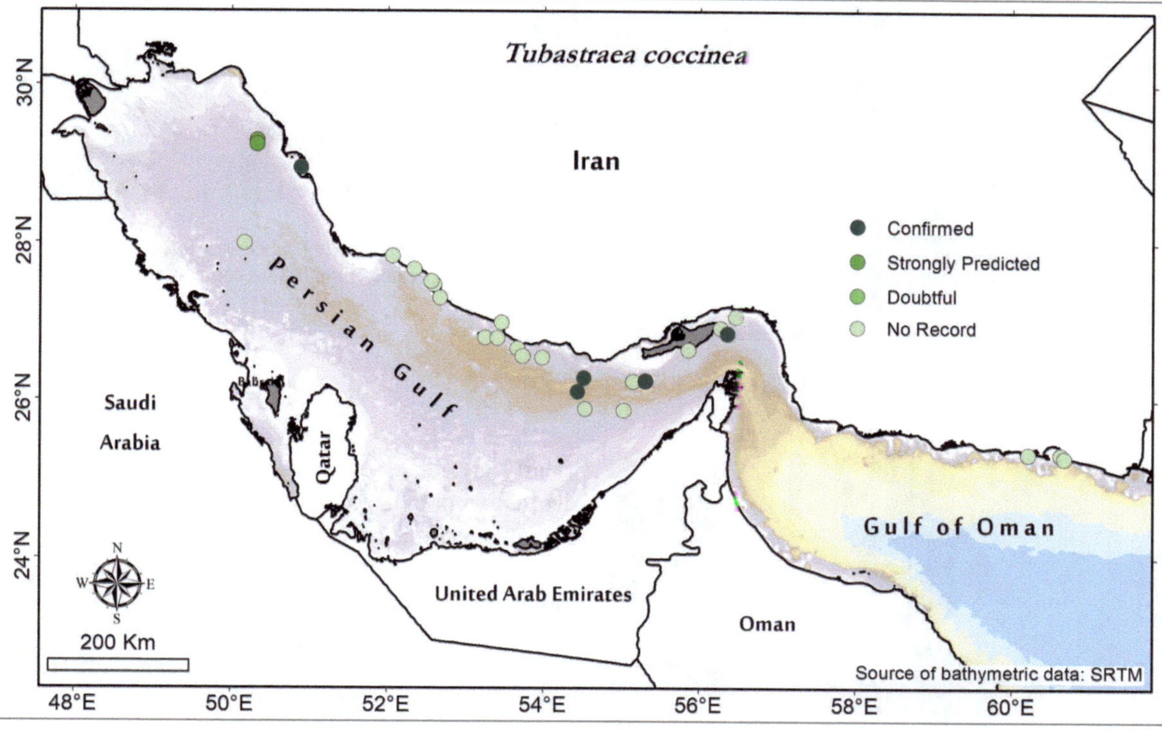

Classification: Phylum—Cnidaria, Class—Hexacorallia, Order—Scleractinia, Family—Dendrophylliidae

Colony form: Colonies of this species exhibit a spherical to mound-shaped morphology, with a maximum diameter of up to 13 cm (Plate 3.67). These colonies exhibit a dendroid growth form, with branches emanating from a relatively narrow base measuring approximately 15 mm in diameter. Each branch, round in cross-section, attains a height of up to 5 cm and bears multiple branchlets. The widest branch tip measures approximately 12 mm.

Corallites: The individual corallites are cylindrical, round and trumpet-shaped. Corallites are characterized by thin walls measuring less than 0.7 mm (Plates 3.68 and 3.69). The calices, reaching depths of up to 8.3 mm, contribute to the overall structure. The septa are arranged in four distinct cycles. The columella, although small, possesses a spongy texture. The wall of the corallites displays patterns and sharp dentations. The arrangement of septa adheres to the plan proposed by Pourtales, with primaries and secondaries extending outward from the wall toward the columella. Importantly, the inner margin of these septa remains intact. In contrast, the third and fourth cycles of septa exhibit dentate features and do not extend significantly beyond the wall. All septa are fenestrate and converge at the base of the columellar fossa. The columella itself presents as a convex, oval structure composed of a spongy tangle of twisted trabeculae.

Colour: Colonies of this species exhibit a striking coloration, ranging from dark to bright red-orange.

Distribution and abundance: Geographically, this coral is documented in both Kuwait, situated in the northwestern region of the Persian Gulf, and Larak Island within the Strait of Hormuz.

3.3.4.5 Genus *Dendrophyllia* de Blainville, 1830

Dendrophyllia colonies exhibit dendroid growth patterns, gradually transitioning into a bushy form through extra-tentacular budding. The corallites are in tubular shape, with septa fused in accordance with the 'Pourtales Plan.' Polyps, are large and fleshy and predominantly active during night. *Dendrophyllia* shares similarities with *Tubastrea*, yet it can be differentiated by the absence of

Plate 3.67 *Tubastraea coccinea* from Greater Tunb. (Photo courtesy of Koosha Dab)

Plate 3.69 Corallite showing calice structures in *Tubastraea coccinea* collected from Greater Tunb. (Photo courtesy of Koosha Dab)

Plate 3.68 Corallites in *Tubastraea coccinea* collected from Greater Tunb. (Photo courtesy of Koosha Dab)

clearly arranged septa in the mature corallites, a feature distinct from the 'Pourtales Plan.' *Dendrophyllia* has been documented in several locations within the northern Persian Gulf and the Gulf of Oman. Specifically, records indicate the presence of *Dendrophyllia* colonies in Kuwait's reefs (Hodgson and Carpenter 1995), and around Larak and Farur islands in the Persian Gulf (Riegl et al. 2012; Shokri and Oladi 2021). Additionally, coral reefs in Chabahar Bay, situated in the Gulf of Oman, have also hosted *Dendrophyllia* species (Orrell 2019). Furthermore, the Sultanate of Oman, within the Gulf of Oman, has been another notable location for *Dendrophyllia* observations (Coles 1996; Claereboudt 2006). Notably, among these documented occurrences, *D. cribrosa* stands as the sole species recorded in Iranian reefs.

Dendrophyllia cribrosa **(Milne Edwards & Haime, 1851)**

IUCN red list assessment		Status in Iran's reefs	
Status	Population trend	Abundance	Population trend
–	–		Unknown

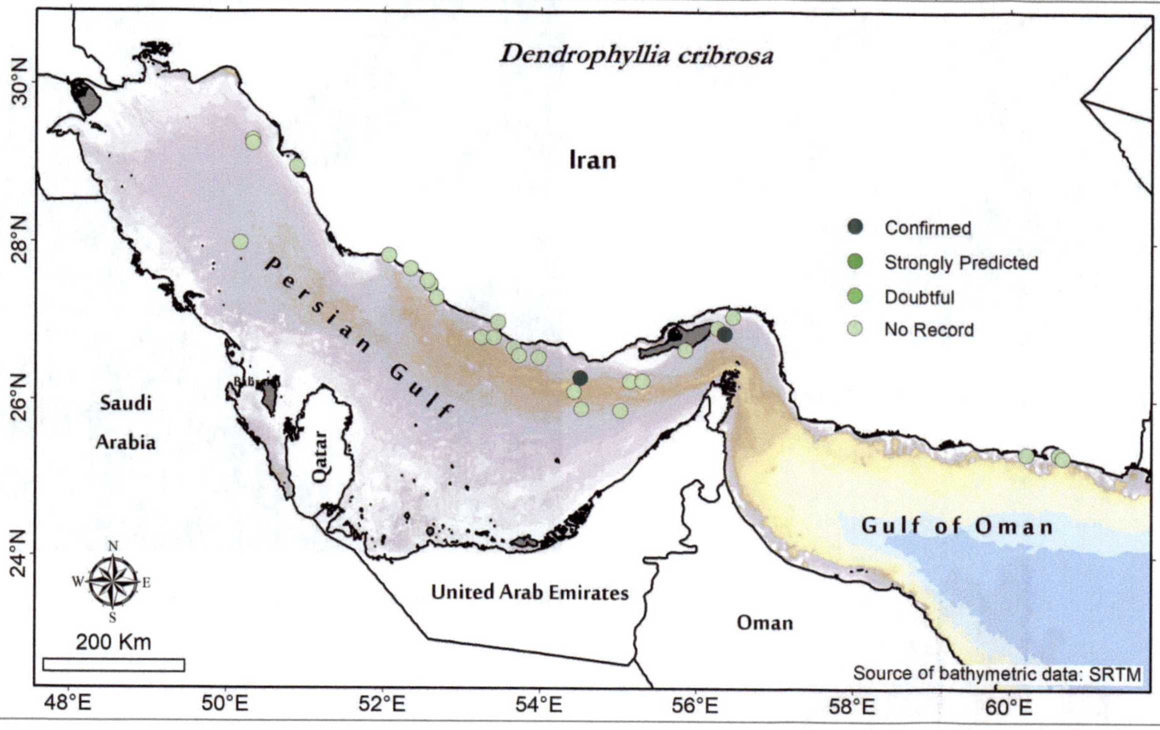

Classification: Phylum—Cnidaria, Class—Hexacorallia, Order—Scleractinia, Family—Dendrophylliidae

Colony form: This species is azooxanthellate and has colonial growth form, characterized by a single robust basal stem from which stout branches emerge (Plates 3.70 and 3.71). These branches, budding from the corallum, spiral around all sides of the large corallum, becoming nearly embedded within its structure (Plate 3.72). These colonies exhibit an upright and tall growth habit, anchored by a sturdy base. The overall corallum assumes a bushy form, with extratentacular branching. Some branches exhibit fusion, while the distinctive oval-shaped corallites possess varying calicular diameters.

Corallites: The corallites, synapticulothecate in nature, display a porous and rough theca (Plates 3.73 and 3.74). Within their architecture, several large axial corallites coexist alongside shorter corallite buds. Well-defined costae traverse the surface, while the septa follow a hexameral arrangement across four distinct cycles, adhering to the established Pourtalès plan. The columella, composed of spongy tissue, contributes to the overall structural integrity. Notably, lateral corallites surpass axial corallites in diameter but remain shorter. The septa are arranged with 38–50 (typically 48) in four cycles.

Colour: Living colonies are enveloped by orange-colored coenosteum tissues.

Distribution and abundance: This species is uncommon and very rare throughout the entire Persian Gulf. In Iran this species was caught in deep trawl net from the depths of 70 m and 67 m off Farur and Larak islands, respectively, in the Persian Gulf.

3.3.5 Family Caryophylliidae

The majority of members within the Caryophylliidae family exhibit an azooxanthellate lifestyle, thriving in the deeper marine realms. While some species in this family form colonies, others remain solitary. Solitary species may be free-living, unattached to hard substrates. The corallum of Caryophylliidae typically is cup-shaped with an elliptical base. The septa, arranged in four to five cycles, exhibit smooth edges. Among scleractinian coral families, the Caryophylliidae stands out as the most diverse family, comprising a total of 301 species (including six infraspecific species) belonging to 43 genera. Notably, three of these

Plate 3.70 *Dendrophyllia cribrosa* from Farur Island

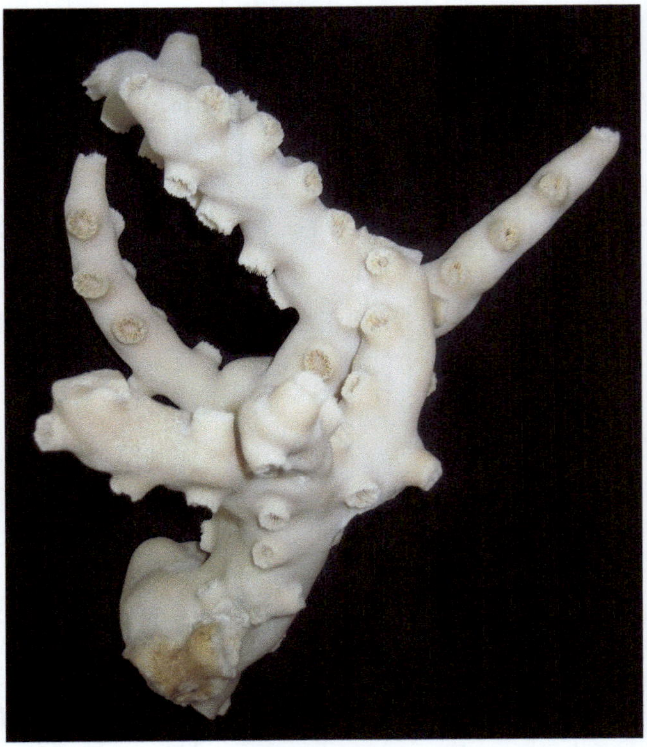

Plate 3.72 Typical branching pattern in *Dendrophyllia cribrosa*

Plate 3.71 *Dendrophyllia cribrosa* from Larak Island

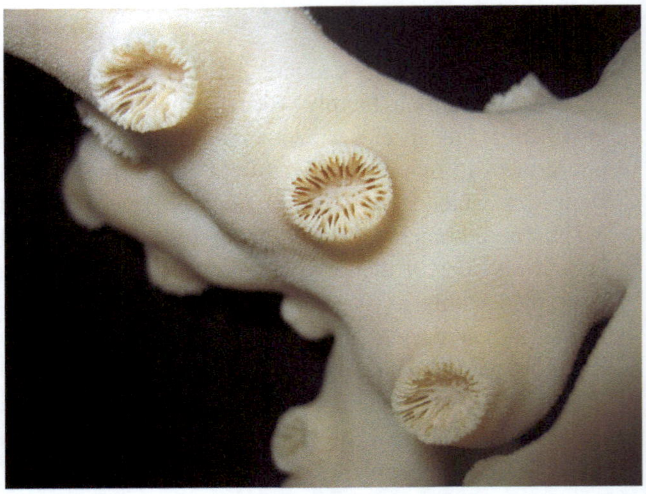

Plate 3.73 Radial corallites in *Dendrophyllia cribrosa*

Plate 3.74 A radial corallite displaying a columella made up of spongy tissue."

genera—*Heterocyathus*, *Paracyathus*, and *Polycyathus*—have been documented in the Persian Gulf and Gulf of Oman (Carpenter et al. 1997; Claereboudt 2006; Ahmed et al. 2023). Specifically, within Iranian reefs, two genera—*Heterocyathus* and *Paracyathus*—each contribute one recorded species.

3.3.5.1 Genus *Heterocyathus* Milne Edwards & Haime, 1848

The early developmental stage of *Heterocyathus* exhibits a solitary lifestyle, initially attaching itself to a small gastropod shell. As growth proceeds, the coral's base gradually encases the entire shell, leaving only an aboral aperture and small lateral respiratory openings. These respiratory passages lead to a spiraling tube, which serves as the habitat for a commensal sipunculid worm *Aspidosiphon corallicola*, which moves the coral across soft substrates, preventing it from becoming buried. *Heterocyathus* species are sometimes zooxanthellate. Tentacles are usually extended only at night. Corals have only one circular corallite and prominent costae.

Heterocyathus aequicostatus is the only species recorded from Kuwait in the Persian Gulf (Carpenter et al. 1997), and Oman's reefs (Claereboudt 2006). Notably, *H. aequicostatus* stands as the sole species documented in the Iranian waters across the Persian Gulf and Gulf of Oman.

Heterocyathus aequicostatus **(Milne Edwards & Haime, 1848)**

Striped bum coral			
IUCN red list assessment		Status in Iran's reefs	
Status	Population trend	Abundance	Population trend
LC	Decreasing		Unknown

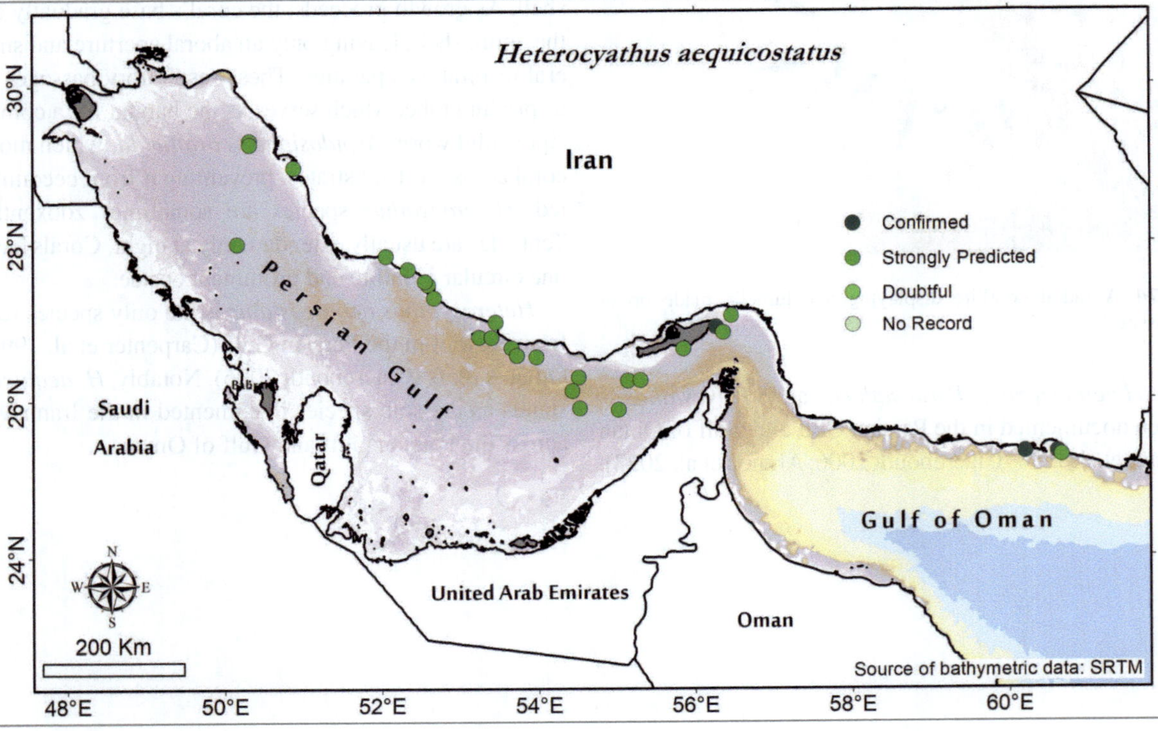

Classification: Phylum—Cnidaria, Class—Hexacorallia, Order—Scleractinia, Family—Caryophylliidae

Colony form: This spcecies exhibits a solitary lifestyle, characterized by their free-living nature and a flat basal structure (Plate 3.75). A notable ecological association exists between corals and spiunculid worms (*Aspidosiphon corallicola*), which is considered commensal (Plate 3.76). In this relationship, the spiunculid worm actively relocates the coral across soft substrates, thereby preventing its burial. During the initial settlement phase, polyp larvae attach themselves to dead micro- molluscs, which subsequently become embedded within the corallum.

Corallites: *Heterocyathus aequicostatus* possesses a single circular corallite, a defining feature of its anatomy (Plate 3.77). These corals exhibit well-defined costae, contributing to their distinctive appearance. Coral polyps extend their tentacles exclusively during night.

Colour: This coral typically displays a pale brown hue, often accompanied by a greenish oral disc.

Distribution and abundance: This species is not common throughout the entire Persian Gulf. In Iran this species is recorded from sandy intertidal areas in the north western Persian Gulf.

3.3.5.2 Genus *Paracyathus* Edwards and Haime, 1848

The genus *Paracyathus* displays a solitary growth habit, firmly attaching itself and tapering toward a pointed base. Notably, the pali fuses with the columella. Within the Persian Gulf, this genus represents an ahermatypic coral, comprising two distinct species: *Paracyathus stokesii* and *Polycyathus marigondoni*. These species have been documented in both Iraq (Ahmed et al. 2023) and Kuwait (Carpenter et al. 1997). However, in Iranian reefs, *P. stokesii* is the recorded species.

Plate 3.75 Lateal view of *Heterocyathus aequicostatus*

Plate 3.76 Aboral view of *Heterocyathus aequicostatus* showing the tiny opening from which a sipunculid worm extends its proboscis

Plate 3.77 Solitary corallite showing calice structures in *Heterocyathus aequicostatus* from Qeshm Island

Paracyathus stokesii (Milne Edwards & Haime, 1848)

Stokes' cup coral			
IUCN red list assessment		Status in Iran's reefs	
Status	Population trend	Abundance	Population trend
–	–		Stable

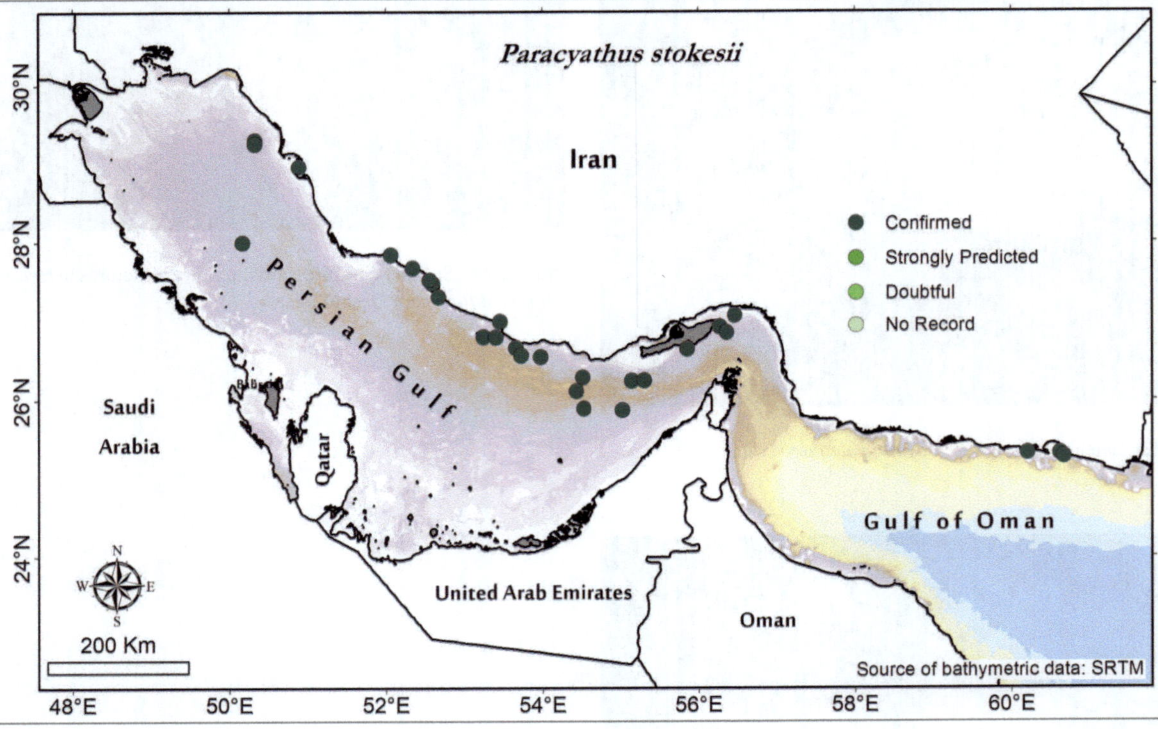

Classification: Phylum—Cnidaria, Class—Hexacorallia, Order—Scleractinia, Family—Caryophylliidae

Colony form: This species within the azooxanthellate category exhibit a solitary growth pattern, tapering toward a pointed base (Plates 3.78 and 3.79). These corals lack an epithecal layer, which typically measures between 1 and 10 mm in height.

Corallites: The calice diameter varies in proportion to the overall size of the corallum, typically falling within the range of 1 to 15 mm (Plate 3.80). Up to 75 septa in four orders that are exsert, featuring granular sides. Notably, the pali is conspicuous and merges with the columella. The columella is narrow and long, measuring approximately 1 mm by 5 mm in mature coral specimens. The costae, lightly granulated, intensify as they approach the septa. Polyps, transparent in appearance, extend primarily during night.

Colour: Underwater, this coral typically displays a pale brown hue. However, the septa and columella of freshly deceased specimens retain a distinctive black coloration, serving as a diagnostic feature.

Distribution and abundance: This species is relatively common throughout the entire Persian Gulf and Gulf of Oman mostly in low-light areas and also in turbid waters.

3.3.6 Family Poritidae Gray, 1840

The Poritidae family comprises a diverse group of distantly related genera. These colonial, hermatypic organisms are predominantly extant. Their colonies exhibit various forms, including massive, laminar, or ramose structures. Corallites within the Poritidae family exhibit a wide size range, although they tend to be compact with minimal coenosteum. The walls and septa of these corals are porous. Within this family, there are four extant hermatypic genera: *Bernardpora*, *Goniopora*, *Porites*, and *Stylaraea*. Notably, *Bernardpora* is a recently established genus (Kitano et al. 2014). Among these, *Porites* stands out as one of the most significant genera on Indo-Pacific coral reefs.

There are some differences in septal arrangement in the coral genera *Porites* and *Goniopora*, highlighting structural

Plate 3.78 *Paracyathus stokesii* from Bushehr waters

Plate 3.80 Solitary corallite showing calice structures in *Paracyathus stokesii*

Plate 3.79 Lateal view of *Paracyathus stokesii*

differences in their corallite anatomy (Fig. 3.3). In *Porites*, the septa are arranged in a simpler pattern with two main cycles, indicated by numbers 1 and 2. The diagram labels key structural elements: (a) the dorsal directive, (b) the ventral triplet, and (c) the lateral pairs. This arrangement is relatively symmetric and shows a basic organization with fewer septa. In contrast, *Goniopora* exhibits a more complex septal pattern with three developmental cycles, numbered 1, 2, and 3. The septa are more numerous and intricately arranged, forming a dense radial pattern typical of *Goniopora* corallites. This distinction in septal complexity is a useful morphological feature for differentiating coral genera in taxonomic studies.

In the context of coral diversity, the Persian Gulf harbors approximately 12 species, distributed across the genera *Goniopora* and *Porites* (Riegl et al. 2012). Conversely, the Gulf of Oman exhibits a more extensive representation, with approximately 19 species spanning three genera: *Goniopora*, *Porites*, and *Stylaraea* (Coles 1996; Claereboudt 2006).

3.3.6.1 Genus *Goniopora* de Blainville, 1830

Gonipora colonies typically exhibit columnar or massive growth forms, although encrusting colonies are also observed. The corallites within *Gonipora* colonies have thick but porous walls, and their calices contain compact septa and a columella. Polyps are elongated, fleshy, and extend both day and night. Interestingly, these polyps are more readily identifiable underwater than when examining their skeletal structures. Each polyp possesses 24 tentacles, and different species exhibit varying polyp shapes and colors, aiding in their underwater recognition. *Gonipora* is closely related to *Porites* and is distinguishable by its long, fleshy polyps that typically extend during daylight hours. Notably, *Gonipora* has the largest corallites within the poritid family and exhibits a distinct

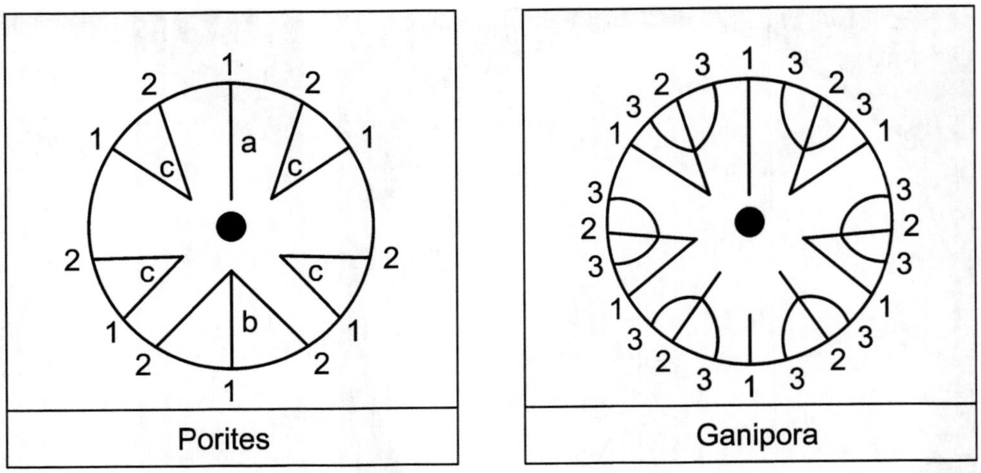

Fig. 3.3 Comparison of septal arrangement between *Goniopora* and *Porites*. (**a**) Dorsal directive, (**b**) Ventral triplet, and (**c**) Lateral pairs. The order of development of successive septal cycles is indicated by the numbers 1, 2, and 3. (Modifid from Bernard, 1905)

septal arrangement. Although septa are generally organized into three orders, their visibility is not consistent (Veron and Pichon 1976). Unlike the fixed septal fusion pattern seen in the genus *Porites*, *Gonipora*'s fusion varies.

In the context of coral diversity, in the Persian Gulf, five species have been recorded for this genus (Riegl et al. 2012). Conversely, the Gulf of Oman exhibits a more extensive representation, with seven species (Claereboudt 2006). In this book four species including *G. lobate*, *G. columna*, *G. djiboutiensis*, and *G. pedunculata* were recorded and described from the Iranian reefs in the Persian Gulf.

Goniopora lobata (Milne Edwards, 1860)

Flower pot true red coral			
IUCN red list assessment		Status in Iran's reefs	
Status	Population trend	Abundance	Population trend
LC	Decreasing		Decreasing

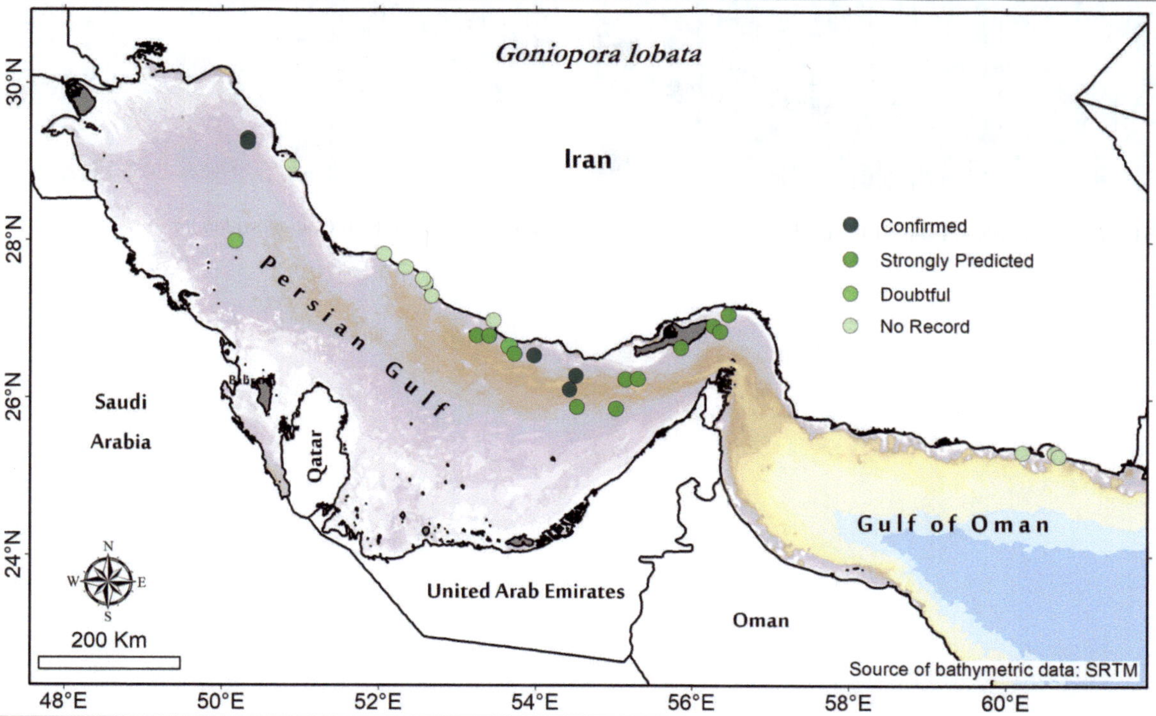

Classification: Phylum—Cnidaria, Class—Hexacorallia, Order—Scleractinia, Family—Poritidae

Colony form: The corallum exhibits columnar growth, with extended polyps typically displaying yellow or light brown hues (Plate 3.81).

Corallites: The polyps attain considerable size, extending several centimeters when fully expanded (Plate 3.82). The corallites exhibit either rounded or polygonal shapes and are relatively shallow, measuring between 2.5 and 4 millimeters in diameter. Within the corallum, the septa—ranging from 18 to 24 in number—display irregular development but predominantly fall into three distinct size classes. Notably, the gonioporoid pattern of fusion is well-developed. First-order septa may bear large spines, resembling pali structures. All septa exhibit variable spine development, while the septal sides appear granulated. Only the two largest septal cycles fuse to form the columella, with the third cycle occasionally remaining abortive and manifesting as a ridge along the theca. The columella itself consists of fused trabeculae and synapticulae. Additionally, the thecae, which encase the corallites, exhibit porosity and measure 1–2 millimeters in thickness.

Colour: The overall coloration ranges from brown to beige.

Distribution and abundance: This particular species appears to be rare, with documented occurrences limited to Kuwait (as reported by Carpenter et al. 1997) and Iran. Within the Iranian reefs, this species has been recorded on the offshore islands (e.g., Farur, Bani Farur, Hendorabi, Kish) toward the Strait of Hormuz in the northern Persian Gulf.

Plate 3.81 *Goniopora lobata* from Kish Island

Plate 3.82 Corallite in *Goniopora lobata*

Goniopora columna (Dana, 1846)

Anemone coral			
IUCN red list assessment		Status in Iran's reefs	
Status	Population trend	Abundance	Population trend
LC	Decreasing		Decreasing

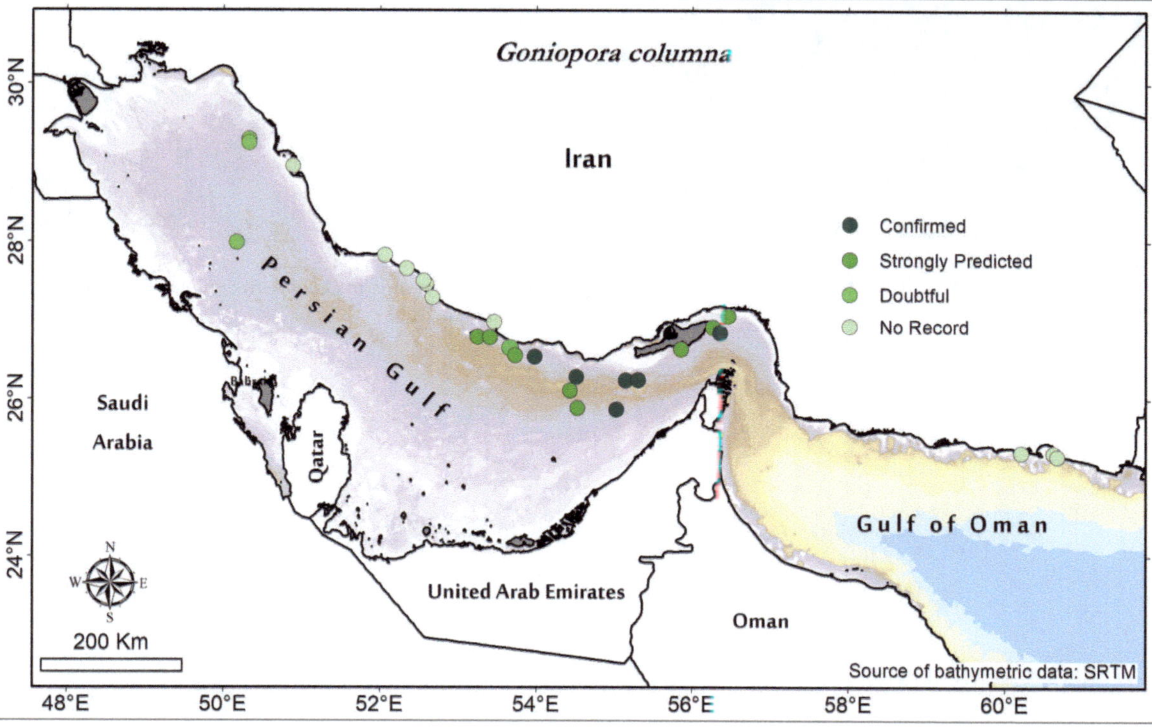

Classification: Phylum—Cnidaria, Class—Hexacorallia, Order—Scleractinia, Family—Poritidae

Colony form: Colonies are characterized by their rounded shape, with uniform corallites and calices exceeding 5 mm in diameter (Plate 3.83). These colonies take up the form of short columns, displaying an oval cross-section. Notably, they predominantly form extensive monospecific stands, particularly in turbid marine environments. Forms large single species stands, especially in turbid water.

Corallites: Corallites positioned near the tops of columns have fine, irregular septa and diffuse columellae (Plate 3.84). Conversely, those situated on the sides of columns exhibit broad, compact columellae and shorter septa. The polyps within these colonies are notably large, characterized by prominent oral cones. The septa within the corallites are exceptionally fine, highly perforated, and irregular.

Colour: Living colonies consistently exhibit uniform coloration, typically ranging from brown to green or yellow with white oral cones. Contracted polyps usually have distinctly different colours.

Distribution and abundance: This species appears to be rare, with documented occurrences limited to Kuwait (as reported by Carpenter et al. 1997) and Iran. Within the Iranian reefs, this species has been recorded on the offshore islands (e.g., Lesser Tunb, Greater Tunb, Abu Musa, Siri, Larak, Hengam) toward the Strait of Hormuz in the northern Persian Gulf.

Plate 3.83 *Goniopora columna* from Lesser Tunb. (Photo credit Koosha Dab)

Plate 3.84 Corallite in *Goniopora columna* from Lesser Tomb (Photo credit Koosha Dab)

Goniopora djiboutiensis (Vaughan, 1907)

Flowerpot coral			
IUCN red list assessment		Status in Iran's reefs	
Status	Population trend	Abundance	Population trend
LC	Decreasing		Decreasing

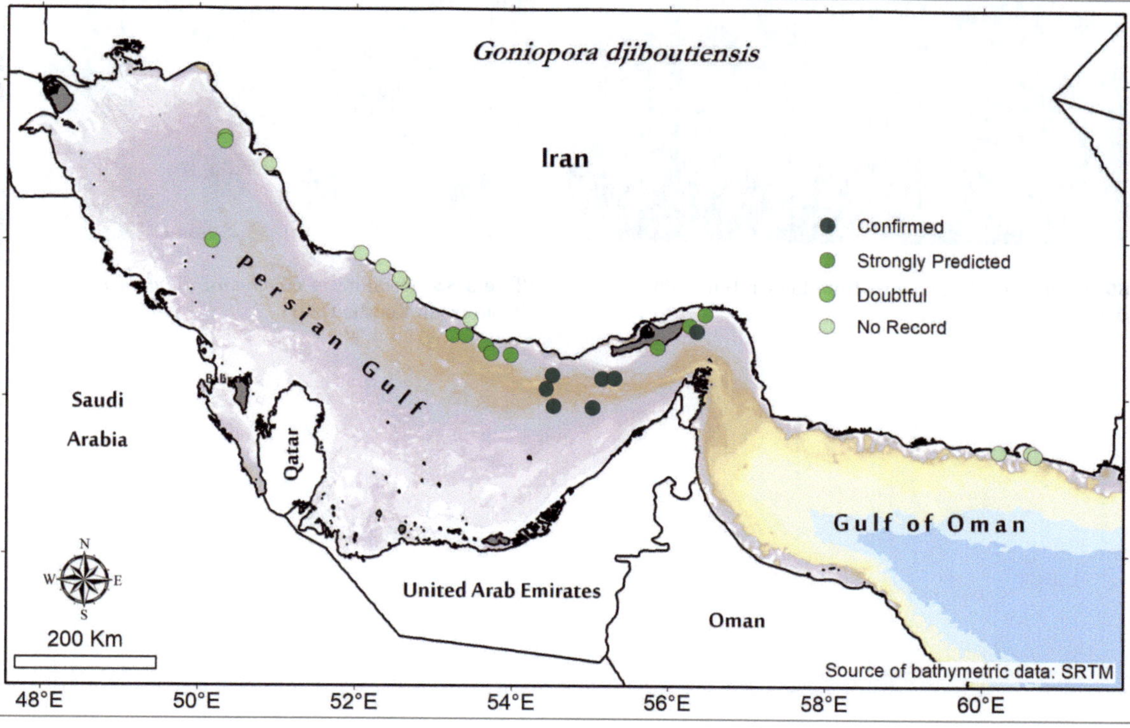

Classification: Phylum—Cnidaria, Class—Hexacorallia, Order—Scleractinia, Family—Poritidae

Colony form: These coral colonies exhibit either a submassive form or short, thick columns, often with an encrusting margin (Plate 3.85). In turbid waters, these corals can form extensive single-species stands.

Corallites: Corallites are prominent, circular to polygonal, porous, and dome-shaped and exhibit extratentacular budding (Plate 3.86). Their diameter measures 5 mm, with a depth of 2–3 mm. The septa are regularly arranged, with 24 septa (up to 36 in larger corallites). Most of the septa follow a gonioporoid arrangement pattern, projecting inwards into the calice to a limited extent. However, 6–8 septa reach the columella. The free margins are serrated, and the sides are highly perforated. At times, the surface is ornamented with clearly visible short protuberances. The wall exhibits a synapticular, fenestrate meshwork with a ragged appearance. Its thickness varies. Columellae are and dome-shaped, and are typically divided into six parts. Each segment follows a deltaic pattern of four septa. The polyps possess large oral cones.

Colour: The color of these corals ranges from pale to dark brown or green. Notably, the oral cones are usually white or blue, which may appear pink in photographs.

Distribution and abundance: This species appears to be rare, with documented occurrences limited to the Iranian offshore islands (e.g., Lesser Tunb, Greater Tunb, Abu Musa, Siri, Farur, Bani Farur, Hengam, Larak, Qeshm) toward the Strait of Hormuz in the northern Persian Gulf.

Plate 3.85 *Goniopora djiboutiensis* from Lesser Tunb. (Photo credit Koosha Dab)

Plate 3.86 Corallite in *Goniopora djiboutiensis* from Lesser Tomb. (Photo credit Koosha Dab)

Goniopora pedunculata (Quoy & Gaimard, 1833)

Lesser Goniopora			
IUCN red list assessment		Status in Iran's reefs	
Status	Population trend	Abundance	Population trend
LC	Decreasing		Decreasing

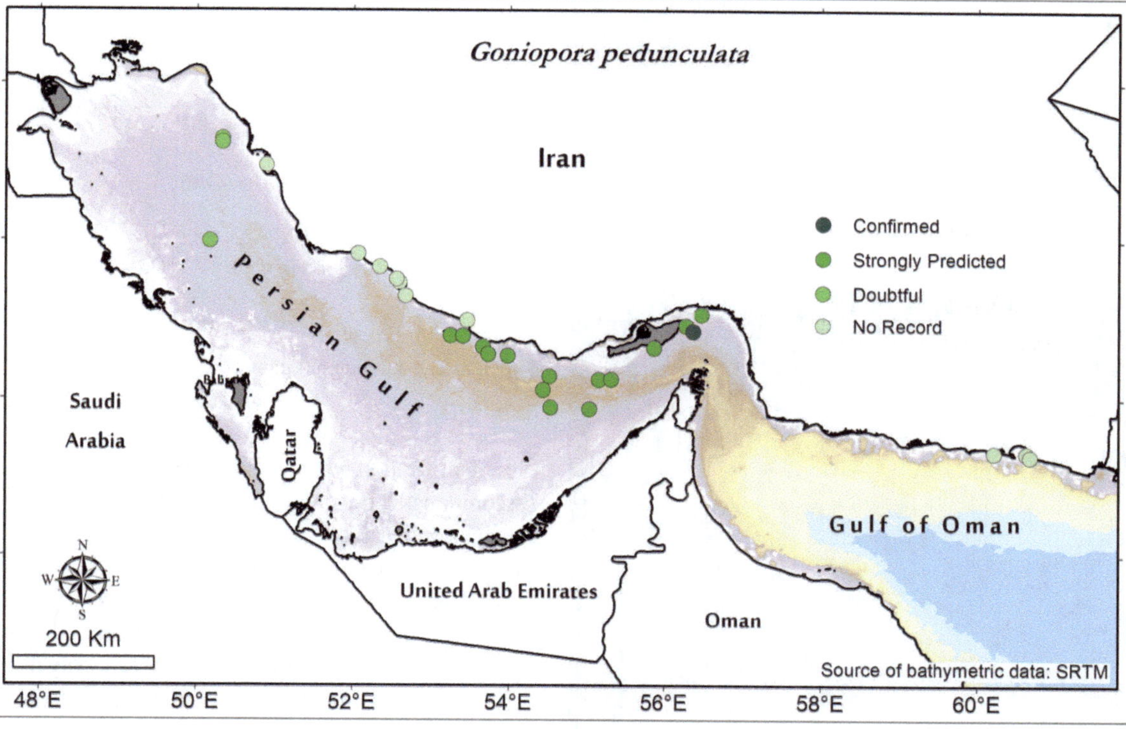

***Classification*:** Phylum—Cnidaria, Class—Hexacorallia, Order—Scleractinia, Family—Poritidae

***Colony form*:** The colony shape occasionally appears hemispherical or encrusting and is predominantly massive, with a maximum diameter of up to 20 centimeters (Plates 3.87 and 3.88).

***Corallites*:** The calices are circular in outline, with thick walls, ranging in size from 2 to 4 millimeters (Plate 3.89). Twelve primary septa are either approximately equal in size or nearly equal, and they exhibit thickness and numerous granules. The upper edges of the septa appear irregularly serrated, with the secondary septa visible in most cases. The columella is well-developed, thick, and closely adjacent or even contiguous, resembling a central crown within the polyp structure.

***Colour*:** Typically brown or green, with a light-colored oral disc when the polyps are expanded. The tentacles exhibit a brown hue, with distinct white tips.

Distribution and abundance: This species is among several other species within the genus *Goniopora* that thrive in environments with turbid water and moderate sedimentation. Although information about its distribution remains incomplete, it has been identified in the Indian Ocean and western Pacific Ocean (Veron 2000) and can be found along the western coast of the Gulf of Aden in Yemen, but its occurrence is exceedingly rare.

Plate 3.87 Expanded polyps in *Goniopora pedunculata* from Larak Island

Plate 3.88 Retracted polyps in *Goniopora pedunculata*

Plate 3.89 Corallites in *Goniopora pedunculata*

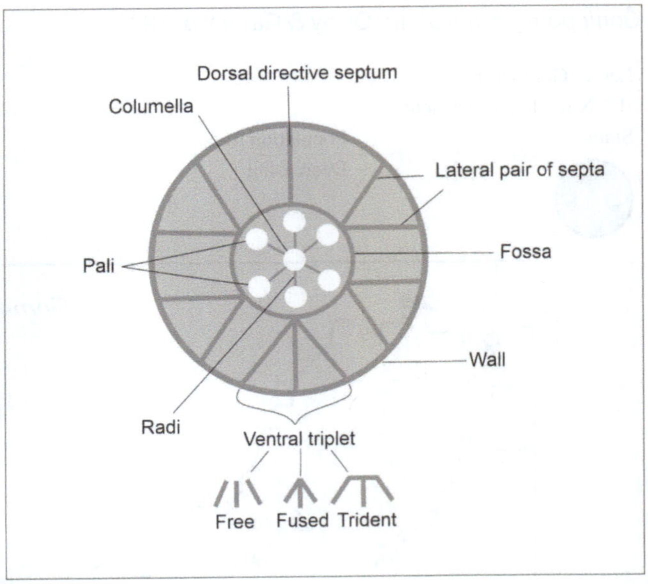

Fig. 3.4 The corallite morphometric characters in *Porites*

3.3.6.2 Genus *Porites* Link, 1807

Porites colonies exhibit diverse growth forms, including encrusting, massive, columnar, branching, or a combination of these structures. When small, massive colonies appear spherical or hemispherical, while larger ones take up a helmet or dome-shaped appearance, often exceeding 5 meters in diameter. The corallites within *Porites* colonies are small, immersed, and possess calices with diameters less than 2 mm (Fig. 3.4). These calices are filled with septa, which bear few denticles and often a palus at their inner margin. Interestingly, *Porites* polyps predominantly extend during nighttime.

Identifying *Porites* species can be challenging due to their variable and small corallites. Key skeletal features used for species identification include: (1) Dorsal Directive Septum: A short septum located dorsally, (2) Lateral Septa: Four pairs of lateral septa, each pair fused at their inner margin (usually larger than the lateral septa in the ventral triplet), (3) Ventral Triplet: Composed of a ventral directive septum (central) and two adjacent lateral septa. The inner margins of these septa may be either free or fused, (4) Pali: Vary in height, absent only in *P. solida*, (5) Denticles: Present on the upper septal margins (Fig. 3.4).

The distinct arrangement of septa—a short dorsal directive septum opposite a ventral triplet—defines this genus (Fig. 3.5). *Porites* is distinguished by a specific pattern of septal arrangement, where a short directive dorsal septum lies opposite a ventral triplet consisting of the ventral directive and two adjacent septa. However, due to the reduction of thecae and septa into arrangements of vertical spines with trabecular linkages, species identification from thin-sections can be challenging, ranging from difficult to nearly impossible. For definitive identification, well-preserved colony surfaces are essential. Additionally, the reduction in skeletal material contributes to significant morphological variability, as patterns of septal fusion can be easily modified. Therefore, the characteristics described below pertain to what is typically observed in the majority of corallites, although substantial variability exists. To achieve a comprehensive understanding, researchers are advised to explore multiple regions within the corallum.

Porites shares similarities with Montipora. However, they differ in growth forms. Corallites in *Porites* are generally larger and more compact, lacking the elaborate thecal and reticulum papillae and tuberculae characteristic of *Montipora*. Additionally, *Porites* corallites contain septa, whereas *Montipora* corallites only have inward projecting septal teeth.

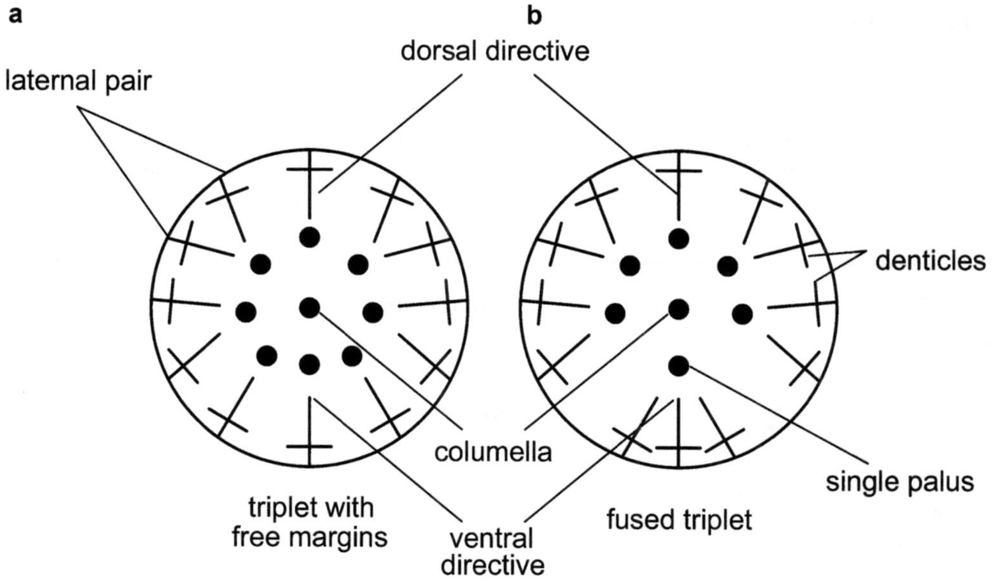

Fig. 3.5 Skeletal features of corallites and fusion patterns of triplets in *Porites*. (**a**) A triplet characterized by free lateral septa and three pali. (**b**) A triplet with lateral and ventral directives fused and one palus. (Modified from Venkataraman and Satyanarayana 2012)

The genus *Porites*, which is prevalent in both the Atlantic and Indo-Pacific regions, plays a pivotal role in Indo-Pacific coral reefs as a primary reef-builder. Its significance dates back to the Eocene epoch. In the context of coral diversity, the Persian Gulf hosts seven recorded species within this genus (Riegl et al. 2012). In contrast, the Gulf of Oman exhibits a more diverse representation, with approximately 11 species (Claereboudt 2006). In this book, five species—namely, *P. harrisoni*, *P. lobata*, *P. lutea*, *P. rus*, and *P. annae*— are documented and described from the Iranian reefs situated in both the Persian Gulf and the Gulf of Oman.

Porites harrisoni **(Veron, 2000)**

Hump coral			
IUCN red list assessment		Status in Iran's reefs	
Status	Population trend	Abundance	Population trend
LC	Decreasing		Stable

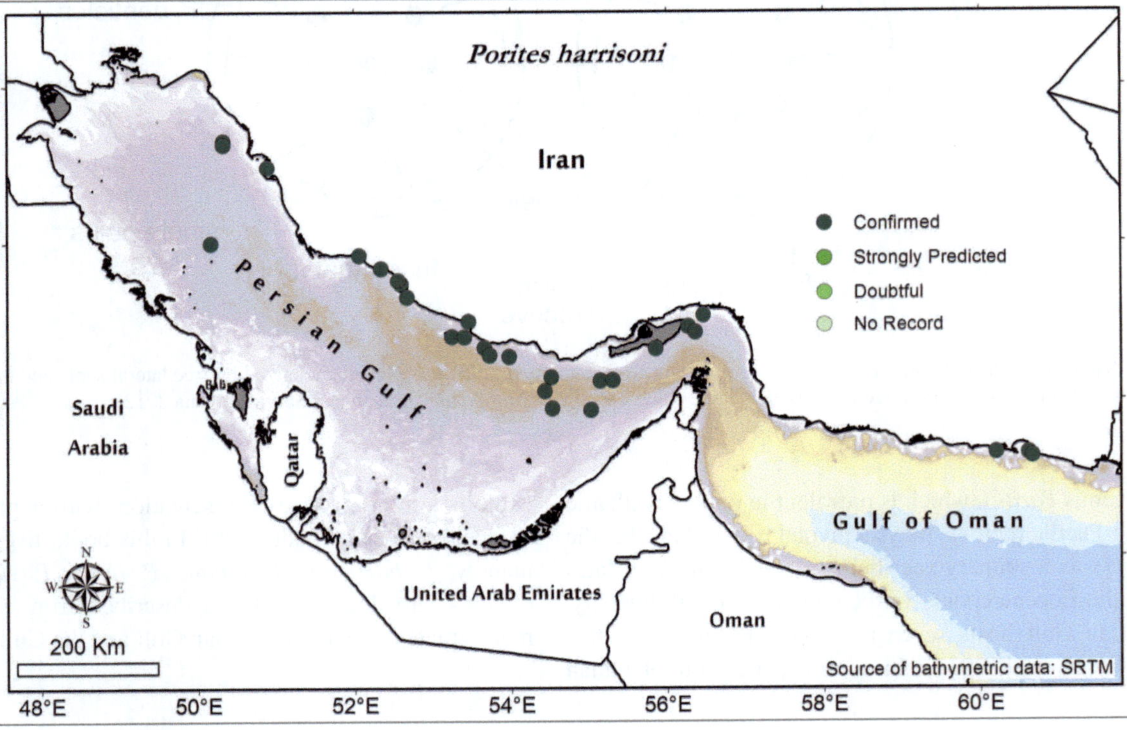

Classification: Phylum—Cnidaria, Class—Hexacorallia, Order—Scleractinia, Family—Poritidae

Colony form: Colonies display a diverse array of growth forms, including submassive, nodular, columnar, and branching structures, and often originate from a broad encrusting base (Plate 3.90). In smaller specimens, they occasionally appear encrusting or foliaceous. However, as they mature, larger colonies develop into distinctive columns with either conical or irregular profiles. Typically, these colonies span less than one meter in diameter and predominantly exhibit vertical growth patterns, reaching heights of up to one meter.

Corallites: Corallites have a diameter of approximately 1–2 mm and exhibit a predominantly polygonal shape (Fig. 3.6; Plate 3.91). Polyps are densely packed and relatively shallow. The septa, although inconspicuous to the naked eye, are small. The spacing between adjacent corallites is typically much less than the diameter of a single corallite. Immature corallites are frequently observed. Commonly, a total of 12 septa are noted. The ventral triplet of septa remains characteristically unfused, although in certain instances, it may fuse due to the presence of a synapticular rod. The other septa are generally of similar size. At the ends of septa and on each septum of the triplet, up to eight large pali are present. If the triplet is fused, only one palus may be observed. Lateral septa are usually free, although some instances show fusion at the palus, resulting in a couplet. The lower synapticular ring (palar ring) is typically complete, while the outer synapticular ring remains incomplete. Typically, there are one or sometimes two denticles on the septa. The septa are linked to the columella through synapticulae. The columella is small and styliform, although its presence is not consistent. Thecae are thin and have a row of spicules along the upper edge. In cases where the theca is wider, it exhibits a porous texture. Notably, Corallum shares similarities with *P. nodifera*, although the latter produces thicker columns and is relatively rare in the region.

Colour: The coloration of *P. harrisoni* typically appears dark brown, occasionally exhibiting shades of gray or blue.

Distribution and abundance: *Porites harrisoni*, a prominent coral species in the Persian Gulf and Gulf of Oman, exhibits distinct morphological features and widespread distribution. It has occasionally been misidentified as *P. compressa*.

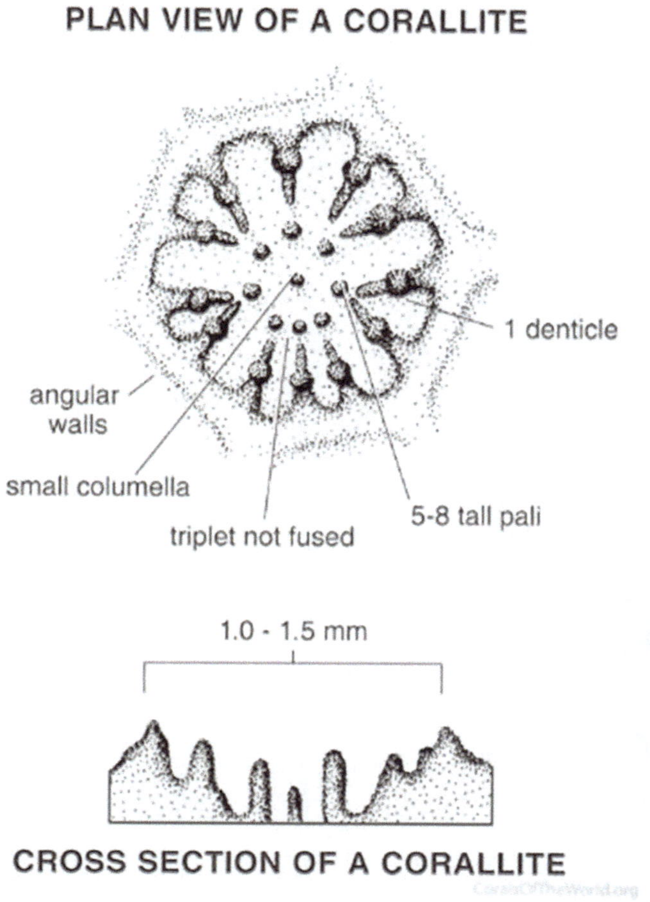

Plate 3.90 Corallite characters in *Porites harrisoni.* (Diagram: Charlie Veron)

Fig. 3.6 *Porites harrisoni* in Larak Island

Plate 3.91 Corallites in *Porites harrisoni*

Porites lobata **(Dana, 1846)**

Lobe coral			
IUCN red list assessment		Status in Iran's reefs	
Status	Population trend	Abundance	Population trend
LC	Decreasing		Stable

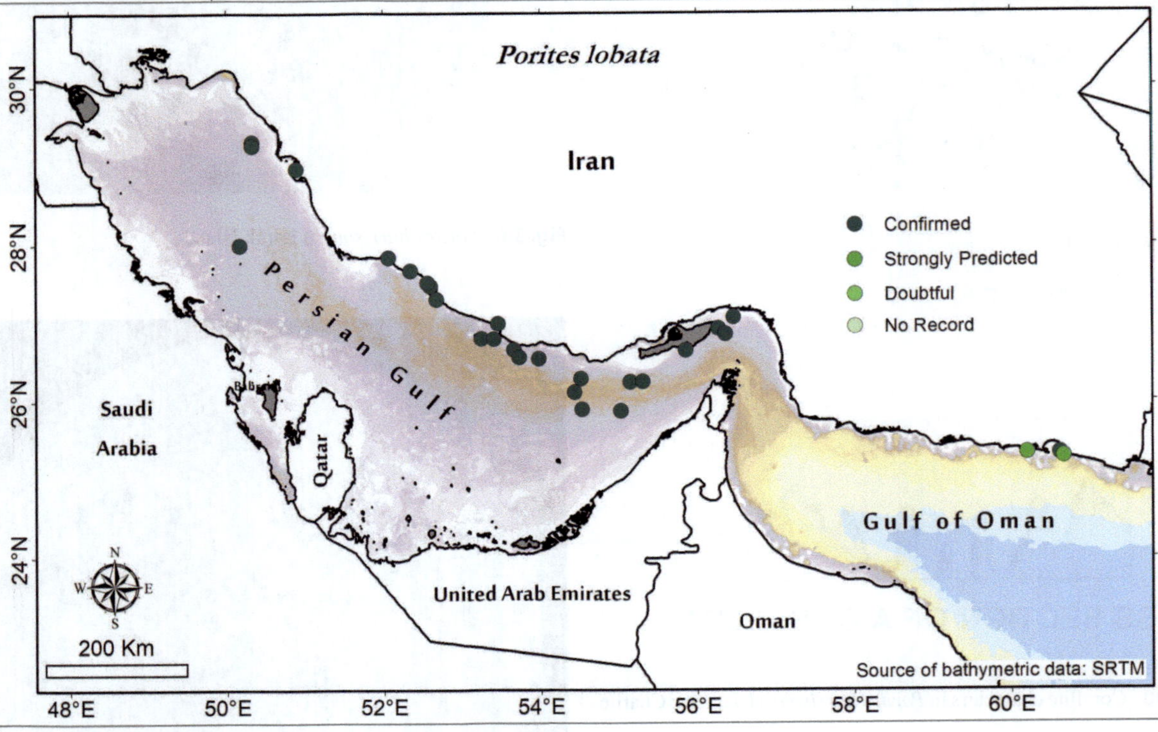

Classification: Phylum—Cnidaria, Class—Hexacorallia, Order—Scleractinia, Family—Poritidae

Colony form: Colonies exhibit a typical hemispherical or helmet-shaped form, can attain considerable dimensions, both in diameter and height, often exceeding 1 meters in diameter (Fig. 3.9; Plate 3.92). The colony surface typically appears smooth. In larger colonies, well-defined columns or lobed protuberances emerge, contributing to their distinctive appearance. Additionally, thick ledges or a series of such ledges encircle the base of larger colonies, sometimes accompanied by surface irregularities or columnar expansions.

Corallites: The calices (individual skeletal cups) within these colonies exhibit an average diameter of approximately 1.5 mm (Fig. 3.7; Plate 3.93). The upper surface of the thick walls consists of three rows of dentides. Specifically, the two outer rows align with the rim of their respective calices, leaving a central row along the summit of the wall. Septa typically bear two denticals between the pali and the outer wall. The columella is consistently present and assumes a mostly round to polygonal shape, particularly in crowded conditions. Notably, the walls of the septa are better developed in this species compared to other local *Porites* species. Despite considerable variability in the overall corallum structure, the septal patterns remain stable, and specimens from the Persian Gulf exhibit remarkable similarity to those from the eastern Pacific (as documented by Glynn et al, 2007). A relatively large dorsal directive septum spans approximately half the length of the lateral pairs. The ventral triplet of septa is characteristically free, with rare fusion between one of the lateral septa and the ventral directive septum. Complete lateral pairs of septa fuse at their tips, forming a distinct and ornamented palus. Occasionally, the ventral directive septum is shorter than the attendant lateral septa in the triplet. A single ornamented spicule is typically found in the middle of the septum. Lateral dentation varies in prominence but is usually present. A well-developed palar ring connects the columella to the directive septa. The thecae (outer skeletal walls) are well-developed and acute, featuring between three and one row of ornamented, large spines. In wider thecae (when corallites are not crowded), the two lateral rows of spines are low, allowing the central row to stand out prominently and align vertically. Spines often fuse distally, forming an acute ridge.

Plate 3.92 *Porites lobata* from Abu Musa Island

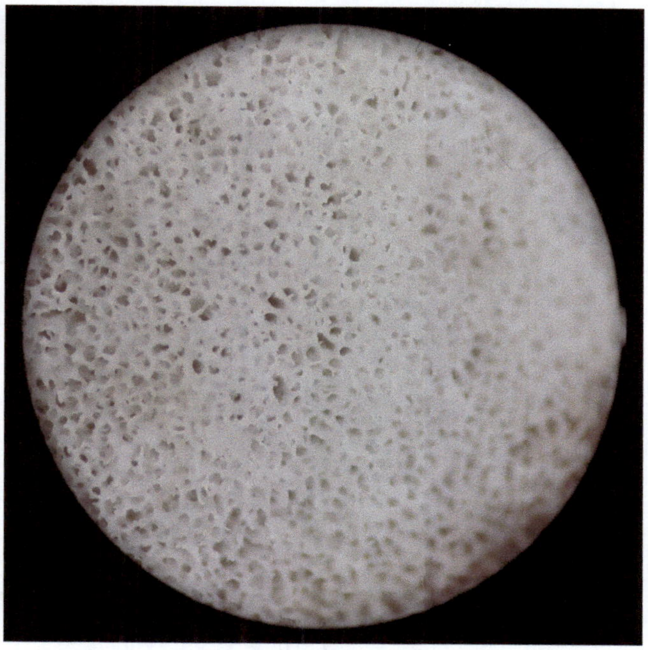

Plate 3.93 Corallites in *Porites lobata*

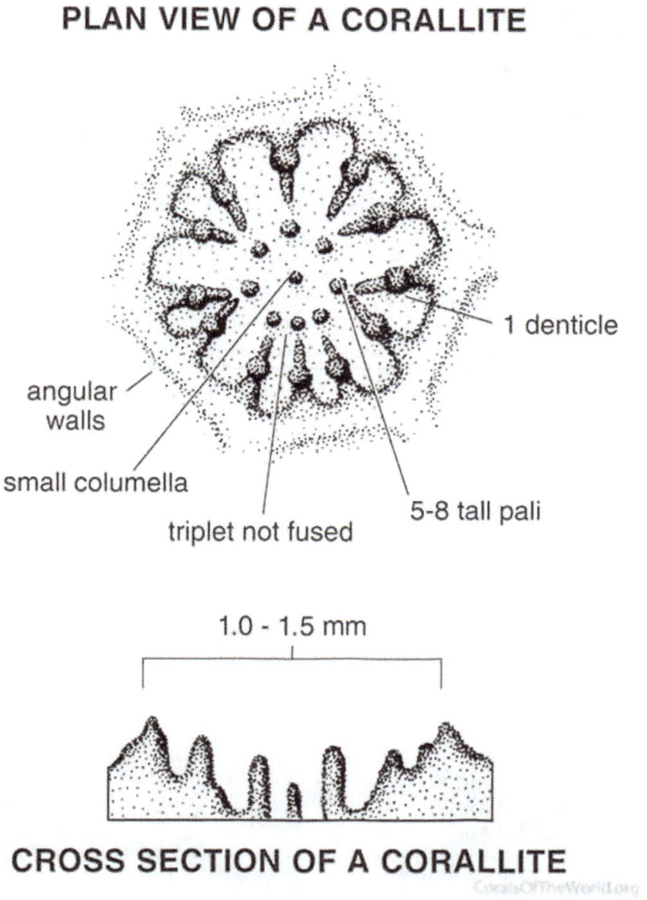

Fig. 3.7 Corallite characters in *Porites lobata.* (Diagram: Charlie Veron)

Colour: Living colonies exhibit uniform colors, typically cream or pale brown, although in shallow water, they may display vivid hues such as bright blue, purple, or green.

Distribution and abundance: This coral species is widely distributed throughout the Persian Gulf and Gulf of Oman. However, in the southern Persian Gulf, it is generally rare (Riegl et al. 2012). It is common in Iranian reefs throughout the Persian Gulf and Gulf of Oman.

Porites lutea (Milne Edwards & Haime, 1851)

Hump coral			
IUCN red list assessment		Status in Iran's reefs	
Status	Population trend	Abundance	Population trend
LC	Decreasing		Stable

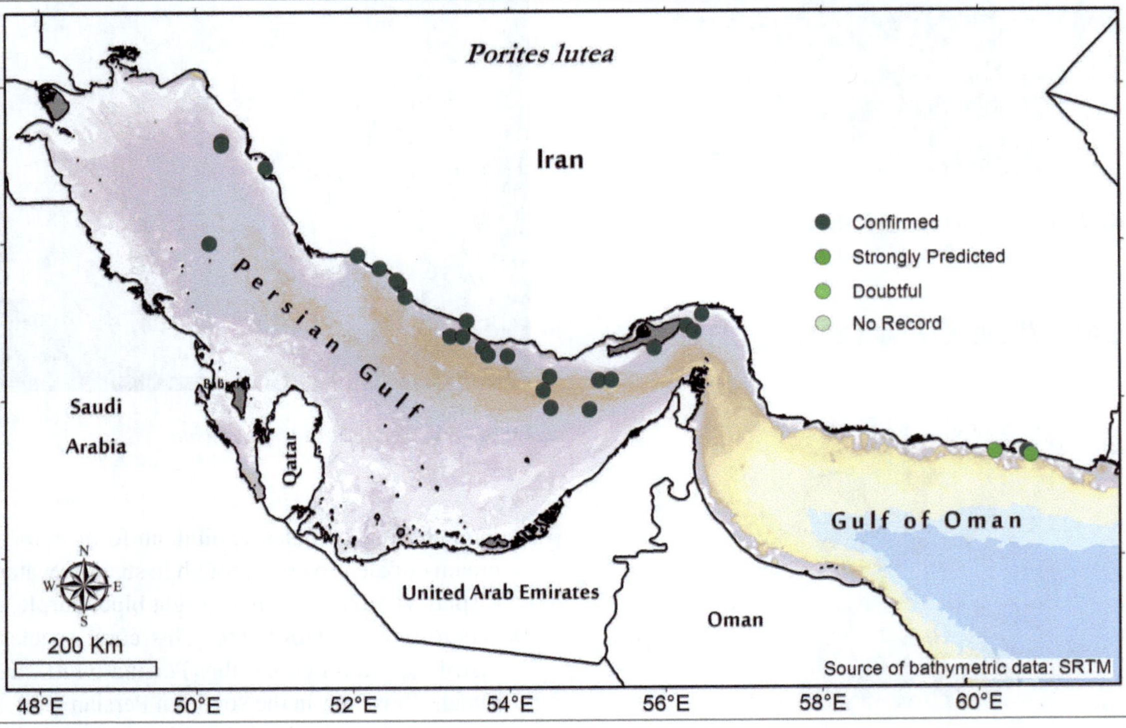

Classification: Phylum—Cnidaria, Class—Hexacorallia, Order—Scleractinia, Family—Poritidae

Colony form: In the Persian Gulf, this coral species exhibits a range of forms, including encrusting, massive, and occasionally (sub) columnar growth (Fig. 3.10; Plate 3.94). Unlike *P. lobata*, it does not develop clear lobes. Colonies typically assume hemispherical or helmet-shaped structures, often achieving considerable size. The colony surface is generally smooth.

Corallites: The calices exhibit an average diameter of approximately 1 mm. Mostly round to polygonal, especially in crowded conditions (Fig. 3.8; Plate 3.95). Fairly shallow, with clearly visible septa. The ventral triplet of septa fuses in a characteristic trident shape (as described by Veron and Pichon, 1976). Five large pali are positioned at the ends of the septa and triplet. If multiple pali are present, the one on the middle directive septum tends to be the largest. Pali connect to the columella via synapticulae (radii), and a lower synapticular ring (palar ring) is complete, while the outer synapticular ring remains incomplete. Septa often bifurcate or become wedge-shaped between the outer synapticular ring and the theca. Only one denticle is observed on the septa. Columella is generally small but mostly present. Absent in some corallites. Thecae is thin, with a row

Plate 3.94 *Porites lutea* from Abu Musa Island

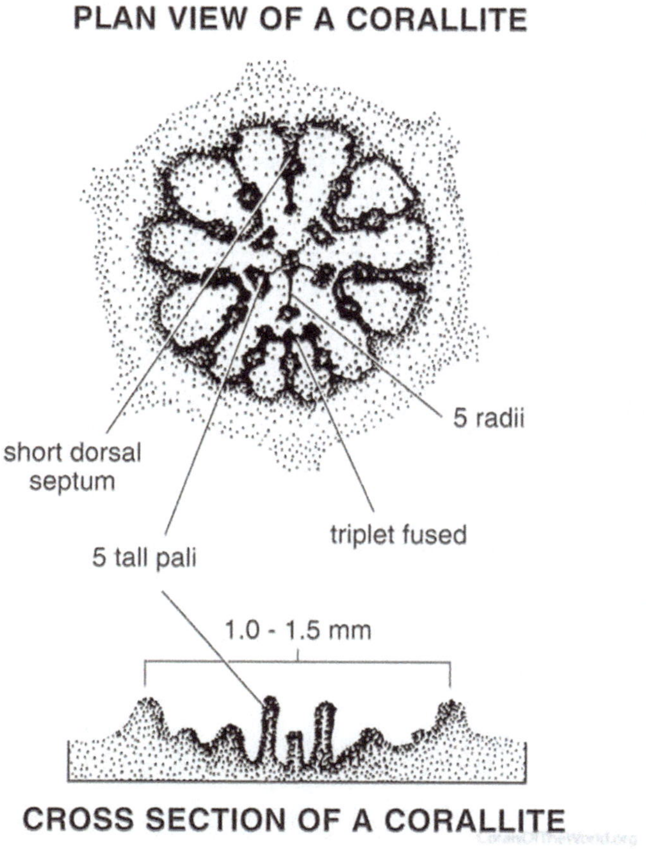

Fig. 3.8 Corallite characters in *Porites lutea.* (Diagram: Charlie Veron)

Plate 3.95 Corallites in *Porites lutea*

of spicules along the upper edge. In wider thecae, a spongy appearance is evident.

Colour: Living colonies typically exhibit cream or yellow hues, although in shallow waters, they may appear brightly colored.

Distribution and abundance: This coral species is common across various habitats throughout the Persian Gulf and Gulf of Oman. It is common in Iranian reefs throughout the Persian Gulf and Gulf of Oman.

Porites rus (Forskål, 1775)

Encrusting Porites			
IUCN red list assessment		Status in Iran's reefs	
Status	Population trend	Abundance	Population trend
LC	Decreasing		Decreasing

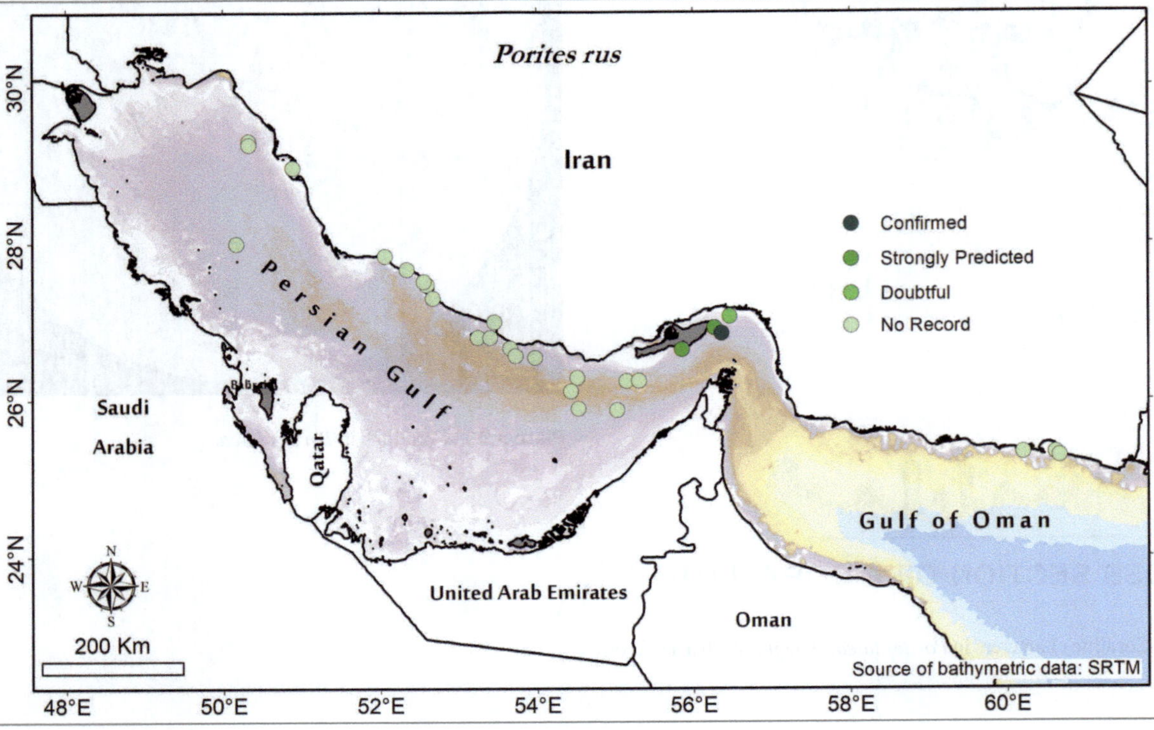

Classification: Phylum—Cnidaria, Class—Hexacorallia, Order—Scleractinia, Family—Poritidae

Colony form: The colony exhibits various forms, including submassive, encrusting or form contorted anastomosing branches (Plate 3.96). These colonies can reach diameters between 20 and 60 centimeters. The colonies thrive in relatively deep areas within the reef.

Corallites: The calyx is typically rounded at its outer boundary, with a very small diameter in the Porites genus, ranging from 0.6 to 0.7 millimeters (Fig. 3.9; Plate 3.97). The corallites are separated by a lattice-like coenosteum, characterized by delicate spines. The arrangement of polyps within the colony follows a distinct series, forming separate short chains. The colony features two primary septa, an outer single septum, and four pairs of lateral septa that merge internally. Additionally, the third mesenteric septum, usually found near the ends, contributes to the overall structure. Granules are present on the upper edge of the septum, and the columella is well-developed, typically consisting of six slender, needle-shaped columns. This morphology is indicative of *Montipora* corals. *Porites rus* can sometimes be mistaken for *Montipora*

Plate 3.96 *Porites rus* from Larak Island

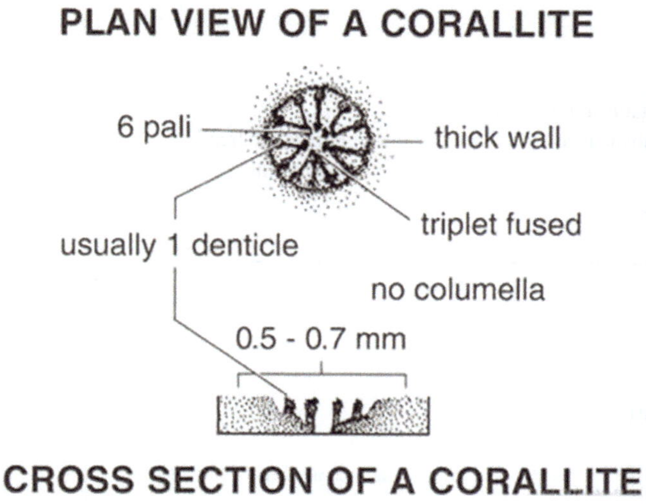

Fig. 3.9 Corallite characters in *Porites rus*. (Diagram: Charlie Veron)

Plate 3.97 Corallites in *Porites rus*

when observed underwater. Overall, this species represents a species complex that likely comprises several cryptic species.

Colour: The color of colonies ranges from pale cream to pale purple and gray. In shallow waters, the branch tips may exhibit brighter colors.

Distribution and abundance: This species is considered an uncommon species in the Persian Gulf and Gulf of Oman. In Iranian reefs, it is found mostly in shallow depths around the offshore islands in the Persian Gulf closer to the Straits of Hormuz.

Porites annae **(Crossland, 1952)**

Encrusting bolder coral			
IUCN red list assessment		Status in Iran's reefs	
Status	Population trend	Abundance	Population trend
LC	Decreasing		Decreasing

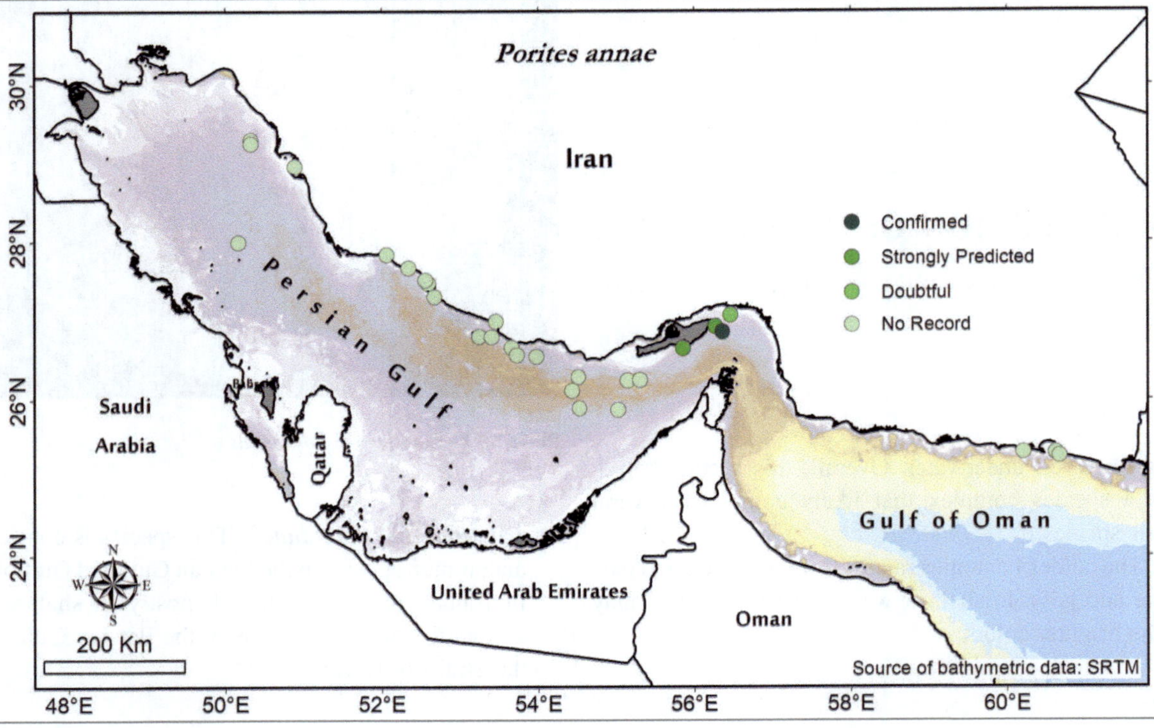

Classification: Phylum—Cnidaria, Class—Hexacorallia, Order—Scleractinia, Family—Poritidae

Colony form: Colonies exhibit nodular anastomosing branches or columns, with bases that can either encrust or be laminar (Plate 3.98). The colony structure develops from a base with an encrusting form, reaching lengths and diameters of approximately 60 centimeters. While the branch tips are typically rounded, some colonies may have nearly cylindrical shapes. Prominent projections emerge from the encrusting base, varying in form from short knobs (reaching only a few millimeters in height) to taller columns (reaching up to 15 centimeters).

Corallites: The polyp structure is relatively deep, with an average size ranging from 1.2 to 1.4 millimeters (Fig. 3.10; Plates 3.99 and 3.100). The corallite walls are thin and serrated. The septal arrangement in *P. anaae* consists of a distinctive pattern, including a single "outer" septum and four pairs of lateral septa (each pair merging at the ends). Additionally, there is usually a counter-directed third mesenteric septum. Occasionally, fusion of the third mesenteric septum occurs in smaller polyp structures, although this is relatively rare. Granules are present on the upper edge of the septum, and the internal septal ridges resemble vertical bars (columella), extending to the upper level of the wall. Overall, the columella is well-developed and horizontally flat.

Colour: The color of colonies ranges from pale to dark green, yellow, purple, or brown. The branch tips or nodules may appear pale or white.

Distribution and abundance: This species is uncommon in the Persian Gulf and Gulf of Oman. In Iranian reefs, it is found mostly in shallow depths around the offshore islands in the Persian Gulf closer to the Straits of Hormuz.

3.3.7 Family Rhizangiidae D'orbigny, 1851

The colonies in Rhizangiidae are typically encrusting and exhibit scattered corallites. Occasionally, they may be solitary. The corallites are tubular and small, with extra tentacles budding. These colonies primarily consist of azooxanthellate species, which do not harbor symbiotic algae (zooxanthellae). Approximately 33 species across four genera have been

Plate 3.98 *Porites annae* from Larak Island

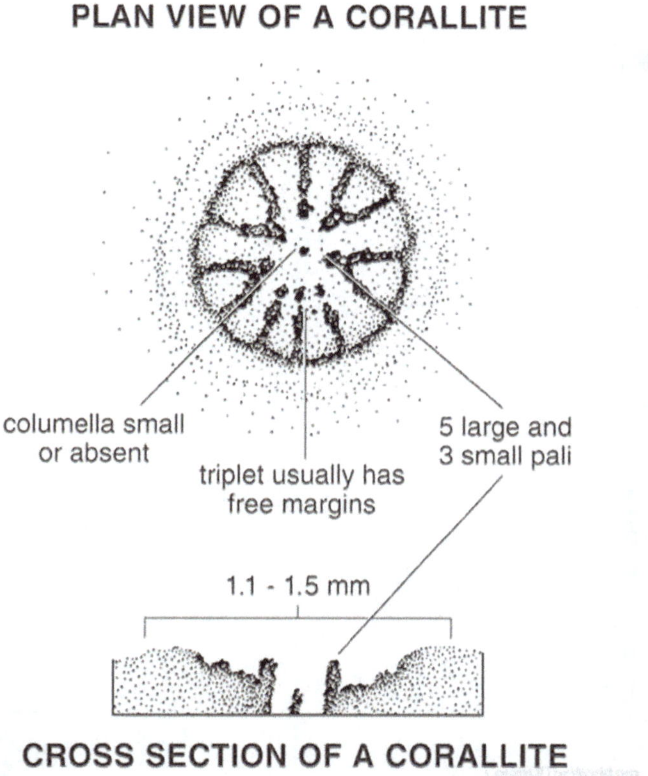

Fig. 3.10 Corallite characters in *Porites annae*. (Diagram: Charlie Veron)

Plate 3.99 Corallites in *Porites annae*

Plate 3.100 Corallites in *Porites annae*

reported. Remarkably, only one species (*Siderastrea savignyana*) from a single genus is recorded from the reefs in the Persian Gulf and Gulf of Oman (Coles 1996; Claereboudt 2006; Riegl et al. 2012).

3.3.7.1 Genus *Siderastrea* Blainville, 1830

Colonies exhibit a boulder-shaped to encrusting growth form. The corallites are cerioid, rounded, or polygonal in shape. Globally, there are five recorded species of Siderastrea, of which one species (*Siderastrea savignyana*) is recorded from the reefs in the Persian Gulf and Gulf of Oman (Coles 1996; Claereboudt 2006; Riegl et al. 2012). The septa within the corallites are evenly spaced and arranged in distinct orders. Columellae, consisting of a few pinnules, contribute to the structural integrity. The walls of the corallites are composed of a network of septa interconnected by cross linkages. Notably, closely resembles *Pseudosiderastrea*.

Siderastrea savignyana (Milne Edwards & Haime, 1849)

African pillow coral			
IUCN red list assessment		Status in Iran's reefs	
Status	Population trend	Abundance	Population trend
LC	Decreasing		Stable

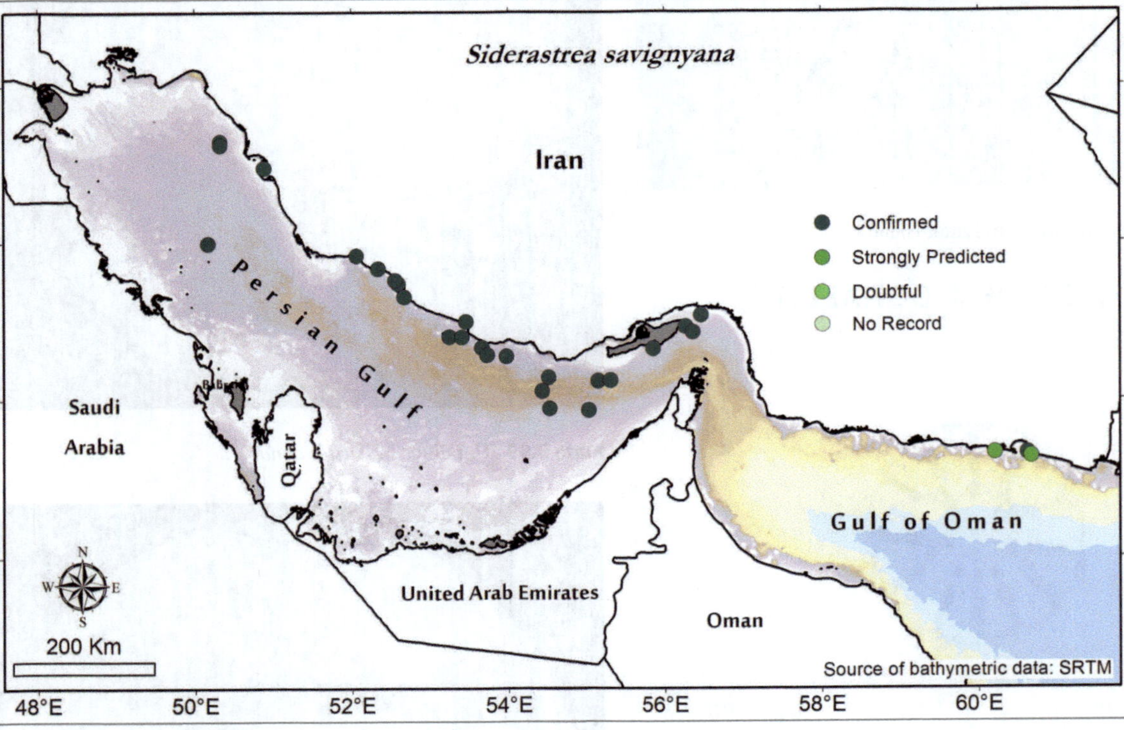

Classification: Phylum—Cnidaria, Class—Hexacorallia, Order—Scleractinia, Family—Rhizangiidae

Colony form: The corallum consists of flat encrusting colonies, often partially buried in sediment or forming massive structures (Plates 3.101 and 3.102). Additionally, it can establish free-living colonies on rubble beds in shallow waters (1–3 meters deep).

Corallites: The cerioid corallites, ranging between 4 and 5 mm in diameter, exhibit a polygonal outline and are not arranged in series (Plates 3.103, 3.104, 3.105, and 3.106). Approximately 40 septa are present, typically fusing together. The septa are neatly arranged, fusing in fan-like groups. The walls of the corallites feature a fine ridge along the top. Importantly, six septa reach the columella without fusing with others. Synapticulae connect all septa in a single row, occurring regularly between the corallite wall and the fusion points of the septa. The septa themselves are compact, and a styliform columella is present.

Colour: The coloration of the walls is typically white, while the polyp color varies between brown and green.

Plate 3.101 *Siderastrea savignyana* from Larak Island

Distribution and abundance: This particular scleractinian coral demonstrates remarkable resilience to the extreme environmental conditions found in the Persian Gulf

Plate 3.102 *Siderastrea savignyana* from Sambarun Bank

Plate 3.103 Corallites in *Siderastrea savignyana* from Larak Island

(Sheppard et al. 1992). This species is commonly found in reef and sandy habitats, but it also occurs in algal beds, where it can be locally abundant.

3.3.8 Family Merulinidae Milne Edwards & Haime, 1857

This diverse group experienced taxonomic changes in 2021 when several genera were reclassified from the Faviidae and Pectiniidae families. This family comprises 24 genera (Veron 2000). Among these, 23 species from nine genera have been documented in the Persian Gulf and Gulf of Oman. All of these genera are extant, hermatypic and colonial. Identifying many species within this group relies heavily on skeletal examination. Most of these corals exhibit encrusting or massive growth forms, characterized by large circular or hexagonal calices and robust septal walls. Additionally, some species display brain-like structures with indistinct calices nestled within sinuous valleys.

Plate 3.104 Detailed view of corallites in *Siderastrea savignyana* from Larak Island

Plate 3.105 Corallites in *Siderastrea savignyana* from Sambarun Bank

Plate 3.106 Detailed view of corallites in *Siderastrea savignyana* from Sambarun Bank

3.3.8.1 Genus *Cyphastrea* Milne Edwards & Haime, 1848

The genus *Cyphastraea* is readily identifiable due to several key features. First, its corallites are typically small, measuring around 3 mm in diameter. These corallites exhibit a mostly round shape and are not densely crowded. Additionally, *Cyphastraea* species display two distinct size-classes of septa. The corallites themselves are plocoid and often well-spaced in many species. The coenosteum is characterized by blistering and may be covered with spinules. The arrangement of septa within *Cyphastraea* is particularly noteworthy. A circular arrangement consists of 12 large septa alternating with 12 smaller septa. These smaller septa are approximately 1/3 to 1/2 the size of the corallite radius and do not extend to the columella—the central column within the corallite. The larger septa are sometimes thickened over the theca (the outer skeletal wall), resulting in a wedge-shaped appearance. Within this genus, two species are represented, including *C. microphthalma* and *C. serailia*, which occurs in the Persian Gulf and the Gulf of Oman (Riegl et al. 2012).

Cyphastrea microphthalma (Lamarck, 1816)

Lesser knob coral			
IUCN red list assessment		Status in Iran's reefs	
Status	Population trend	Abundance	Population trend
LC	Decreasing		Stable

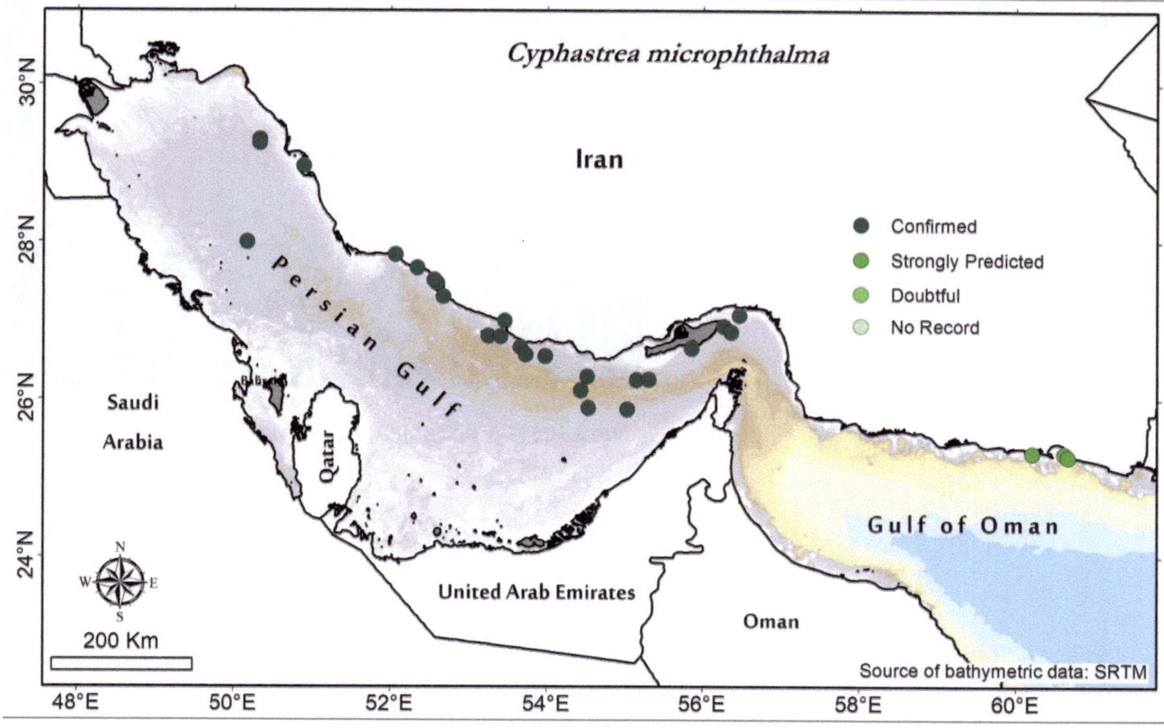

Classification: Phylum—Cnidaria, Class—Hexacorallia, Order—Scleractinia, Family—Merulinidae

Colony form: This species exhibits a massive growth form, ranging from hemispherical to columnar (Plate 3.107). Typically, colonies of this species encrust surfaces, although occasionally they can form massive structures. These colonies are often infested with bivalves, especially Lithophaga species, and polychaete worms.

Corallites: Corallites are plocoid, well-spaced, approximately round, with a diameter of up to 4 mm (Plate 3.108). The thecae can vary, either appearing thick and well-exsert or confluent with the coenosteum. Extratentacular budding is observed in this species. In most specimens, there are 10 first-order septa—an unusual number for hexacorallians. This diagnostic feature is easily verified in the field. The first-order septa are notably larger and thicker, forming a distinct single cycle. Paliform lobes are present on the septa of this cycle, sometimes taking the form of one or two twisted dentations. These first 10 septa extend to the columella. In contrast, second-order septa are reduced to thin lamellae along the corallite wall. Septal sides of all orders exhibit strong granulation. The columella is well-developed and trabecular. The costae are exsert, with first-order costae being larger than second-order ones. The coenosteum is covered with ornamented spines.

Colour: The overall coloration of the coenosarc is typically pale and uniform whitish, occasionally with a darker oral field. In turbid or shaded habitats they may appear dark brown or dark green.

Distribution and abundance: This coral species is common and regularly distributed throughout the Persian Gulf and Gulf of Oman, making it one of the most distinctive and abundant corals across various habitats.

Plate 3.107 *Cyphastrea microphthalma* from Kharg Island

Plate 3.108 Corallites in *Cyphastrea microphthalma*

Cyphastrea serailia **(Forskål, 1775)**

Lesser knob coral			
IUCN red list assessment		Status in Iran's reefs	
Status	Population trend	Abundance	Population trend
LC	Decreasing		Stable

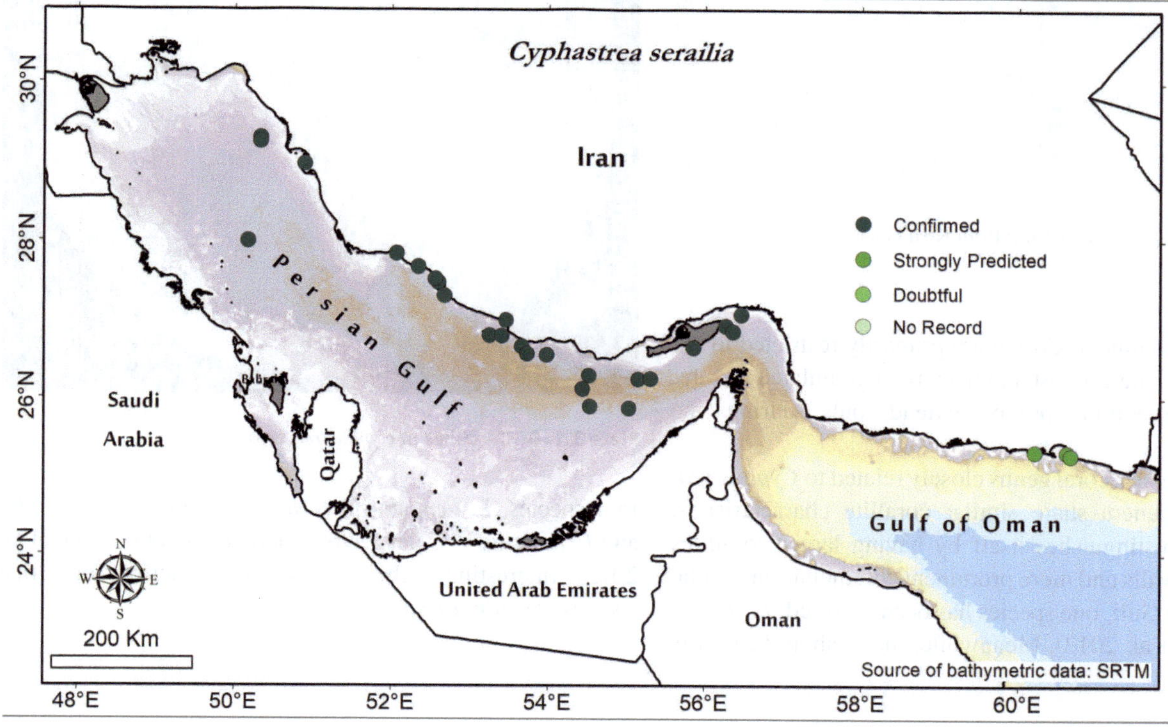

Classification: Phylum—Cnidaria, Class—Hexacorallia, Order—Scleractinia, Family—Merulinidae

Colony form: This species exhibits a massive growth form, ranging from hemispherical to columnar, or even encrusting (Plate 3.109).

Corallites: Its corallites are generally round and can vary in exsertion, ranging from sub-cerioid to plocoid, with diameters up to approximately 4 mm (Plate 3.110). Calices are typically 1.5–2.8 mm in diameter. The septa are organized into two very unequal orders, each consisting of 12 septa. The first-order septa are markedly larger and thicker, forming two cycles of different sizes. Paliform lobes are present on the septa of this cycle, along with prominent dentations. Importantly, the first 12 septa of the larger cycle extend to the columella. In contrast, the second-order septa are reduced, lack paliform lobes, and do not reach the columella. Septal sides of all orders exhibit strong granulation. The columella is usually inconspicuous but well-developed and trabecular. The costae are either equal or subequal, often poorly developed, and heavily ornamented with granulated perithecal spines. These costae do not alternate strongly. The thecae vary significantly in height and thickness. They may be abruptly delineated from the coenosteum or be completely confluent. The coenosteum often appears blistered and is consistently covered with granulated exothecal spines.

Colour: The overall coloration of the coenosarc is typically whitish, occasionally with a darker oral field.

Distribution and abundance: This coral species is common and regularly distributed throughout the Persian Gulf and Gulf of Oman, making it one of the most distinctive and abundant corals across various habitats.

3.3.8.2 Genus *Echinopora* Lamarck, 1816

Echinopora colonies exhibit a diverse array of forms, including massive, arborescent, or foliaceous structures, sometimes combining these morphologies. The corallites are typically plocoid, featuring calices that can reach up to 5 mm in diameter. Septa are exsert and irregular, while columellae

Plate 3.109 *Cyphastrea serailia* from Kish Island

Plate 3.110 Corallites in *Cyphastrea serailia*

are usually prominent. Costae are primarily restricted to the corallite wall. The coenosteum displays a granulated texture. Notably, Echinopora polyps extend only during the nighttime.

Echinopora is a coral genus closely related to *Cyphastrea*. While both genera share similar corallite characteristics, *Echinopora* distinguishes itself by having larger corallites with thicker walls and more prominent columellae and septa. In the Persian Gulf, one species has been reported: *E. irregularis* (Riegl et al. 2012). Meanwhile, the Arabian Sea hosts four species, *E. gemmacea, E. hirsutissima, E. grandicula*, and *E. lamellosa* (Coles 1996; Claereboudt 2006; Riegl et al. 2012). Interestingly, the present study describes only one species, "*E. hirsutissima*".

Echinopora hirsutissima (Milne Edwards & Haime, 1849)

Prickly-pored coral			
IUCN red list assessment		Status in Iran's reefs	
Status	Population trend	Abundance	Population trend
LC	Decreasing		Decreasing

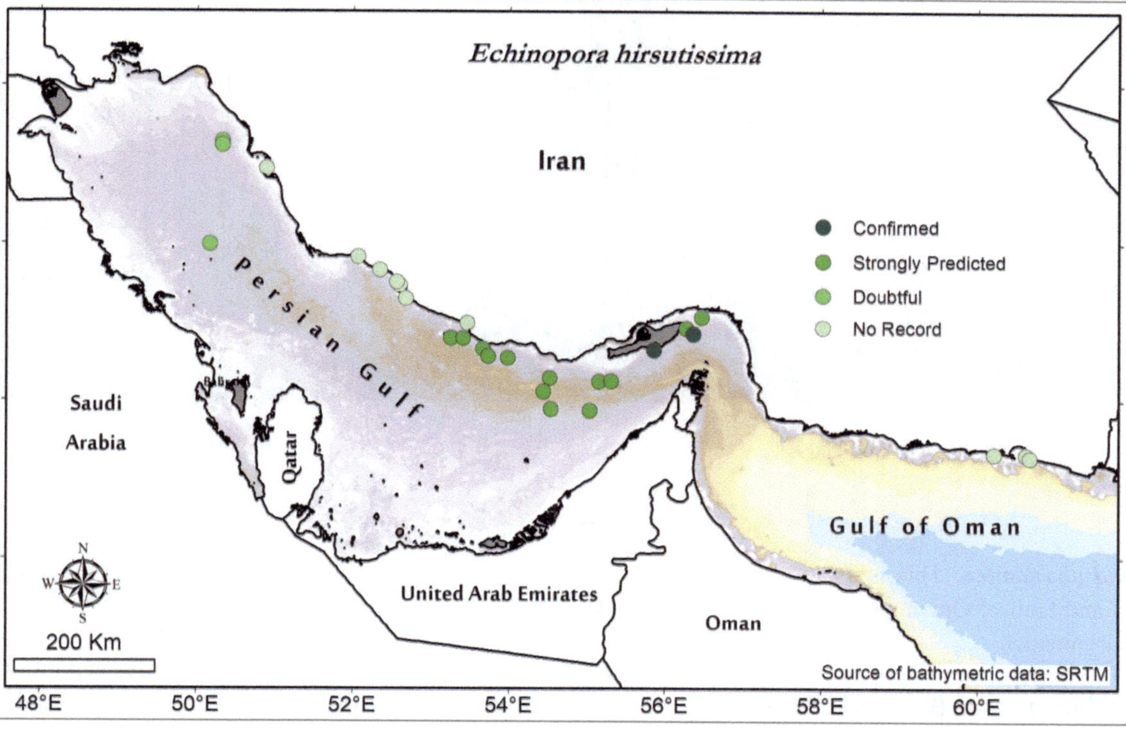

Classification: Phylum—Cnidaria, Class—Hexacorallia, Order—Scleractinia, Family—Merulinidae

Colony form: Colonies exhibit various growth forms, including submassive or encrusting structures, as well as laminar explanate or contorted shapes that rise vertically into hollow projections without apertures at their tops (Plate 3.111).

Corallites: The corallites, measuring 4–7 millimeters in diameter, are superficially or slightly protruding, occasionally inclined, and characterized by thick walls and prominent skeletal features (Plate 3.112). Notably, the larger septa are fairly thick, highly exsert, and extend to the columellae. These septa typically possess two or three very prominent lobes, including an outer septal lobe that may appear flattened or divided, sometimes resembling a replacement for the first costal spine. The well-developed costae exhibit strong beading and are either equal or subequal, adorned with numerous tall spines ornamented by abundant granulations and echinulations. The coenosteum is densely covered with thick, finely elaborated spinules. *Echinopora hirsutissima*, shares similarities with *E. gemmacea*, although it exhibits less coarse corallite structures and smaller corallites.

Plate 3.111 *Echinopora hirsutissima* from Larak Island

Colour: The coloration of these colonies varies, ranging from mustard to green, brown, or purple, often accompanied by pale septo-costae.

Plate 3.112 Corralites in *Echinopora hirsutissima*

Distribution and abundance: This is a rare species in the Persian Gulf and Gulf of Oman, primarily found in shallow reef environments.

3.3.8.3 Genus *Dipsastraea* Blainville, 1830

Dipsastraea is a common genus of large polyp stony (LPS) corals with a worldwide distribution. Key traits associated with *Dipsastraea* include (Chevalier 1971): these corals typically form massive colonies that can reach substantial sizes, colonies are typically massive, exhibiting either flat or dome-shaped growth forms, and corallites are monocentric and plocoid, projecting slightly above the colony surface, each with its wall. Generally, well separated, although occasional fusion may occur (similar to the closely related genus *Favites*). Budding primarily occurs intra-tentacularly and is mono- to tri-stomodeal (resulting in one to three new mouths). Paliform lobes are commonly formed by simple or compound trabeculae. Additionally, multitrabecular paliform lobes can develop from a uniform fan system. Polyps extend primarily at night and have a simple circle of tapering tentacles, often with pigmented tips. *Dipsastraea* shares similarities with *Favites*, but the latter has cerioid corallites. The distinction between the two genera can be somewhat arbitrary, as *Dipsastraea* corallites tend to subdivide equally, while *Favites* corallites usually subdivide unequally, resulting in daughter corallites of different sizes. *Dipsastraea* is likely the most common coral genus in the Persian Gulf, Gulf of Oman and possibly even the Indian Ocean (Riegl et al. 2012). These corals can be found in various habitats within the Persian Gulf, including marginal areas.

In the Persian Gulf and Gulf of Oman, the coral genus *Dipsastraea* is represented by six distinct species: *D. pallida*, *D. speciosa*, *D. favus*, *D. rotumana*, *D. matthaii*, and *D. amicorum* (Coles 1996; Claereboudt 2006; Riegl et al. 2012). Notably, five species—*D. favus*, *D. speciosa*, *D. matthaii*, *D. pallida*, *D. rotumana* —have been documented in Iranian reefs across both regions.

Dipsastraea favus (Forskål, 1775)

Head coral			
IUCN red list assessment		Status in Iran's reefs	
Status	Population trend	Abundance	Population trend
VU	Decreasing		Decreasing

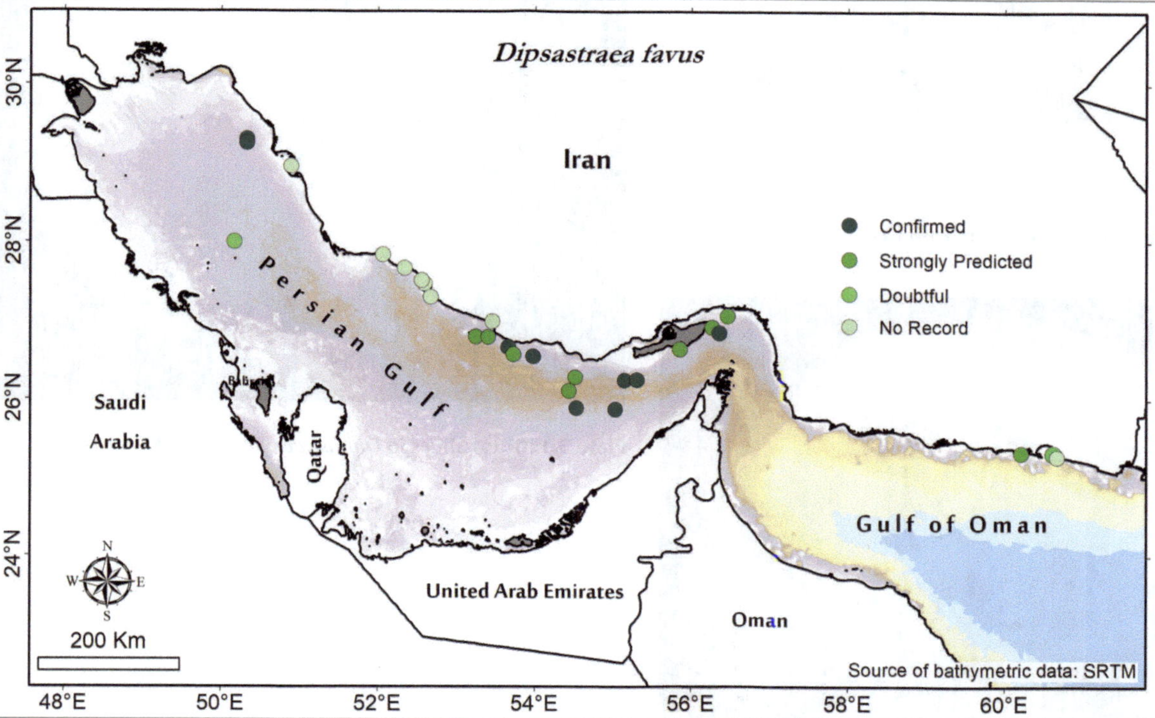

Classification: Phylum—Cnidaria, Class—Hexacorallia, Order—Scleractinia, Family—Merulinidae

Colony form: The colonies are massive and rounded, occasionally flattened (Plates 3.113 and 3.114). The most common growth form is hemispherical, but flat colonies also occur.

Corallites: Corallites are plocoid and have a diameter of 12–20 mm and can extend up to 5 mm above the colony surface (Plates 3.115 and 3.116). They are conical, often protruding, and usually circular but may show irregular shapes. Endotheca (inner wall) depth varies, with some corallites having deep cylindrical calices. Septa are not consistently ordered and typically project outward. Paliform lobes are sometimes present, although they may be barely recognizable or form a distinct crown. Costae aligns between adjacent corallites. Coenosteum is usually blistered.

Colour: *Dipsastraea favus* exhibits highly variable colors, ranging from yellow to reddish hues, often with mottled grey and red patches.

Plate 3.113 *Dipsastraea favus* from Larak Island

Distribution and abundance: These corals are very common and regularly distributed throughout the entire Persian Gulf and Gulf of Oman, thriving in various habitats.

Plate 3.114 *Dipsastraea favus* from Kish Island

Plate 3.115 Corallites in *Dipsastraea favus* from Larak Island

Plate 3.116 Corallites in *Dipsastraea favus* from Kish Island

Dipsastraea speciosa (Dana, 1846)

Larger knob coral			
IUCN red list assessment		Status in Iran's reefs	
Status	Population trend	Abundance	Population trend
LC	Decreasing		Stable

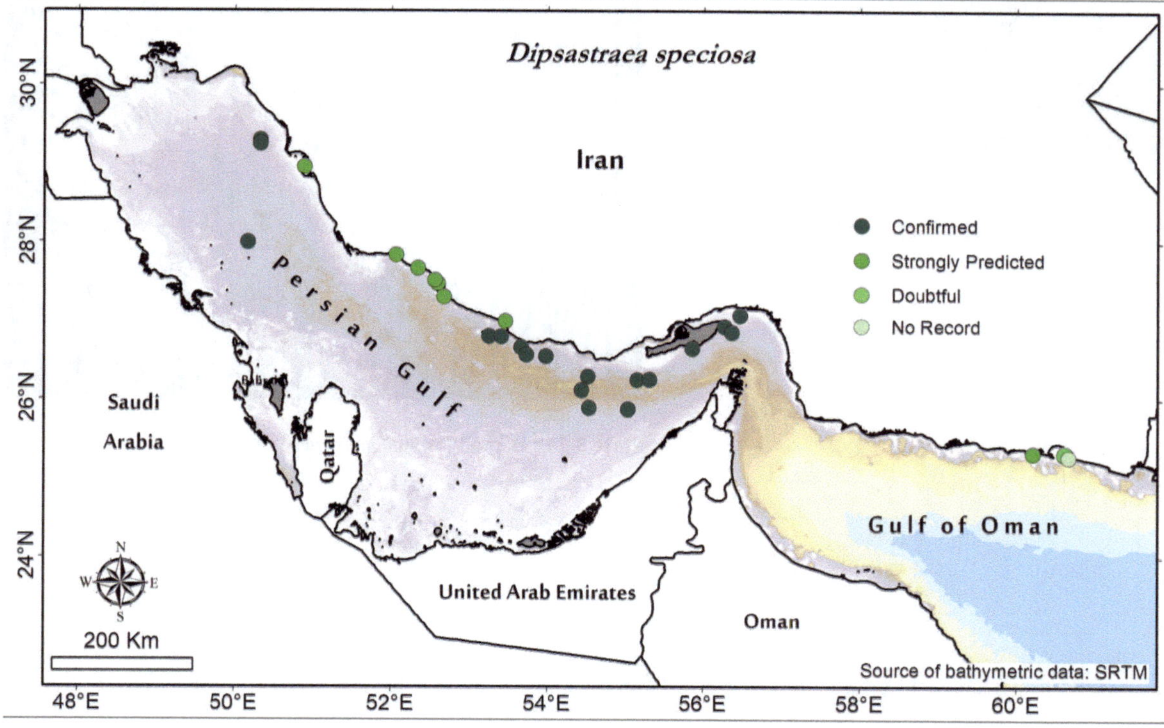

Classification: Phylum—Cnidaria, Class—Hexacorallia, Order—Scleractinia, Family—Merulinidae

Colony form: Colonies exhibit massive hemispherical growth forms, clearly displaying plocoid features (Plate 3.117).

Corallites: Corallites are plocoid, with diameters ranging from approximately 6–10 mm (usually around 7 mm) (Plate 3.118). They are distinctly exsert. Septa are arranged in 2–3 orders. The first-order septa consistently reach the columella, while second-order septa occasionally do so. Third-order septa are reduced or abortive. Similar to *D. pallida*, the septa are thin and regularly exsert. Fragile dentations occur on the peripheral parts of septa, while irregular dentations characterize their inner portions. Paliform lobes or spines are not always present and are generally less conspicuous than in *D. pallida*. The columella is spongy, composed of twisted and fused trabeculae. However, in some colonies, the columella may be poorly developed. Theca (outer wall) thickness is notable. While similar to *D. pallida*, this coral can be distinguished by its circular or elliptical calices, which are less crowded and shallower.

Colour: The color of *D. speciose* ranges from reddish to orange, often with a lighter oral field.

Distribution and abundance: Distribution is uncommon but likely extends throughout the entire Persian Gulf and Gulf of Oman, across various habitats.

Plate 3.117 *Dipsastraea speciosa* from Sambarun Bank

Plate 3.118 Corallites in *Dipsastraea speciosa*

Dipsastraea matthaii (Vaughan, 1918)

IUCN red list assessment		Status in Iran's reefs	
Status	Population trend	Abundance	Population trend
LC	Decreasing		Stable

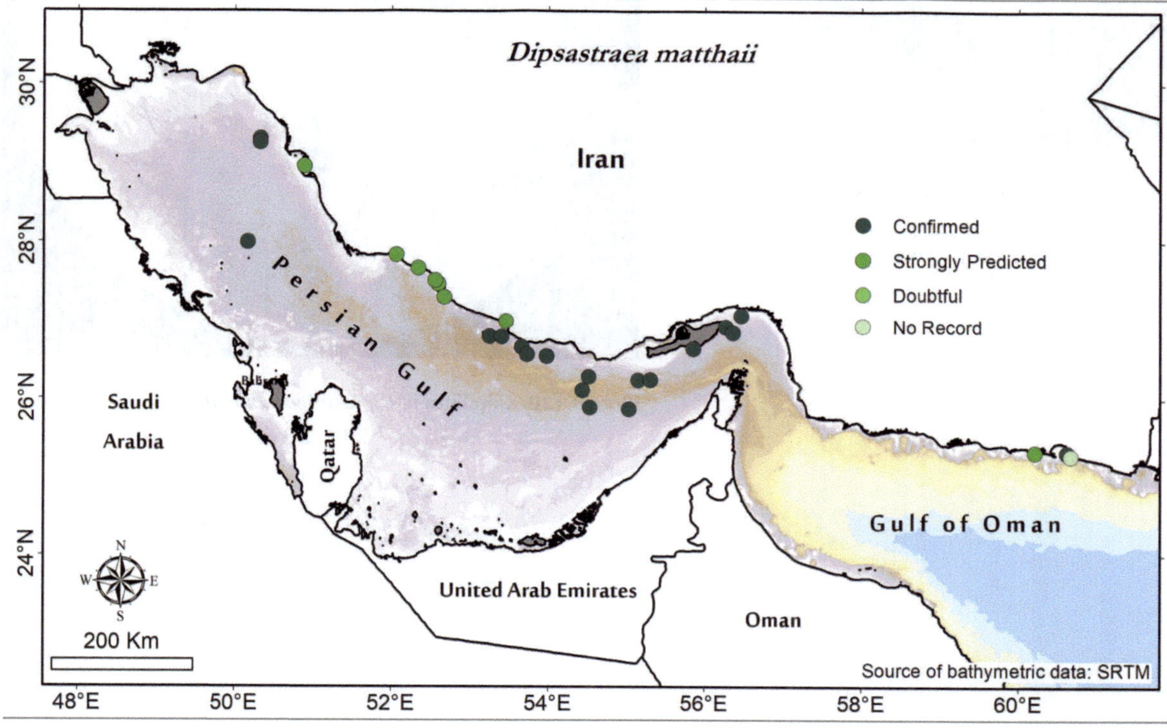

Classification: Phylum—Cnidaria, Class—Hexacorallia, Order—Scleractinia, Family—Merulinidae

Colony form: Colonies exhibit massive and rounded growth forms, occasionally appearing flat or encrusting (Plate 3.119).

Corallites: Corallites are crowded, plocoid, and circular, with diameters typically around 6–10 mm (Plate 3.120). They have shallow, open calices. Septa are thickened and exsert, featuring large teeth near the wall. A well-developed paliform crown is present, and sometimes three cycles of septa form a series of concentric crowns. Costae are beaded and usually not joined with the costae of adjacent corallites. The coenosteum is usually blistered. Similar to *D. pallida*, budding in *D. matthai* occurs entirely intratentacular. However, *D. matthai* has shallower, more open calices with more exsert septa. The paliform crown or series of crowns in *D. matthai* is more prominent, forming concentric circles of elongated dentations that are readily recognizable underwater. In contrast, *D. favus* has markedly larger and more protruding corallites than *D. matthai*, along with substantially different calicular structures.

Colour: The color varies from yellow to different shades of brown, occasionally appearing gray or pale green.

Distribution and abundance: Distribution is relatively common throughout the entire Persian Gulf and Gulf of Oman.

Plate 3.119 *Dipsastraea matthaii* from Abu Musa Island

Plate 3.120 Corallites in *Dipsastraea matthaii*

Dipsastraea pallida (Dana, 1846)

Knob coral			
IUCN red list assessment		Status in Iran's reefs	
Status	Population trend	Abundance	Population trend
LC	Decreasing		Stable

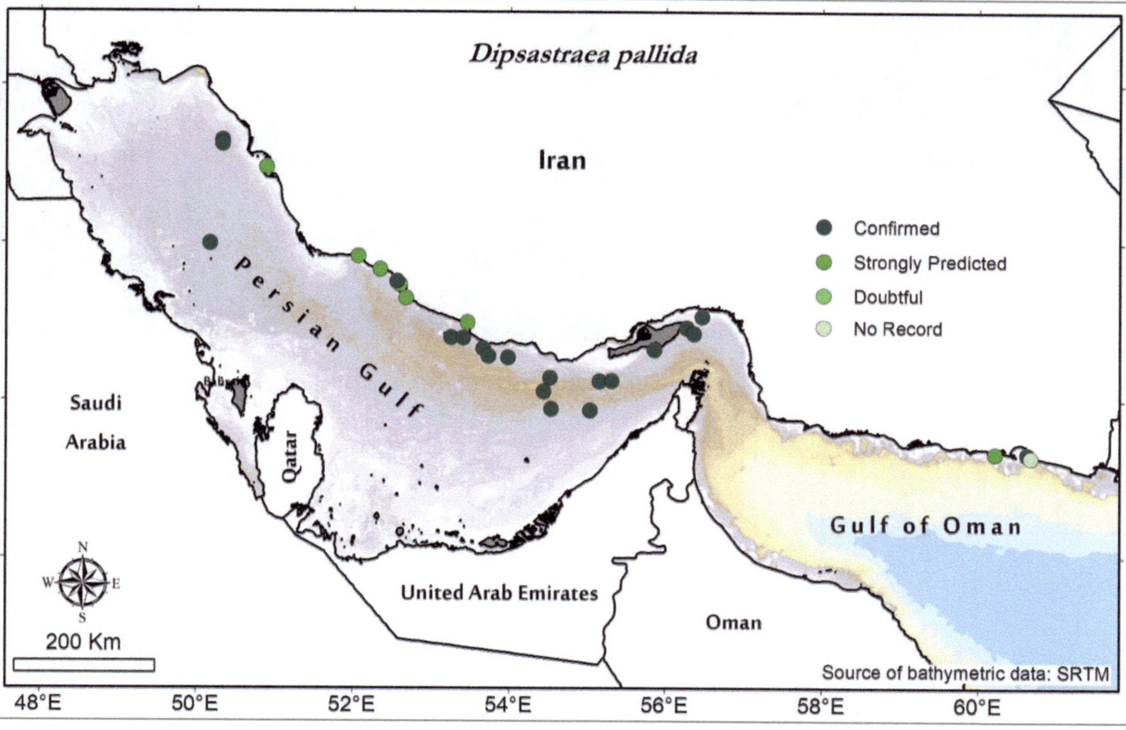

Classification: Phylum—Cnidaria, Class—Hexacorallia, Order—Scleractinia, Family—Merulinidae

Colony form**:** Colonies are massive and rounded, exhibiting a plocoid or ploco-cerioid growth form (Plate 3.121).

Corallites: Corallites have a diameter of 6–10 mm and are usually less than or equal to 2 mm above the colony surface (Plate 3.122). The most common growth form is hemispherical, although flat colonies can also occur. Corallites in *D. pallida* are smaller than those in *Favia favus* (8–12 mm). They are less conical and more tubular, often appearing crowded and polygonal. Septa project outward and are thickened over the theca. They exhibit small, regular dentations and a well-developed paliform crown. Some septal sides are also dentate with fine granulations. Septa are irregularly arranged, mostly in 2–3 orders. The first-order septa consistently reach the columella, while second-order septa may vary in length. Similar to *F. favus*, septa descend steeply into the calyx from the periphery toward the center. Calices are unequal, with spacing typically around 2.5–3.5 mm. In general, calices are less spaced than in *F. favus*, and they can be deep (up to 9 mm) with an average diameter of approximately 8 mm. Paliform lobes are frequently developed on most first-order septa, but not necessarily on higher-order septa. Costae are well developed and can be differentiated into first and second orders, which may vary in size. The columella is always present, usually dense and spongy, consisting of twisted and fused trabeculae. Its development can vary. Theca is thick and the coenosteum (skeleton) is smooth or slightly blistered.

Colour: Colonies exhibits highly variable colors, ranging from yellow to reddish hues, often with mottled grey and red patterns.

Distribution and abundance: These corals are very common and regularly distributed throughout the entire Persian Gulf and Gulf of Oman, thriving in various habitats.

Plate 3.121 *Dipsastraea pallida* from Kharg Island

Plate 3.122 Corallites in *Dipsastraea pallida*

Dipsastraea rotumana **(Gardiner, 1899)**

IUCN red list assessment		Status in Iran's reefs	
Status	Population trend	Abundance	Population trend
LC	Decreasing		Decreasing

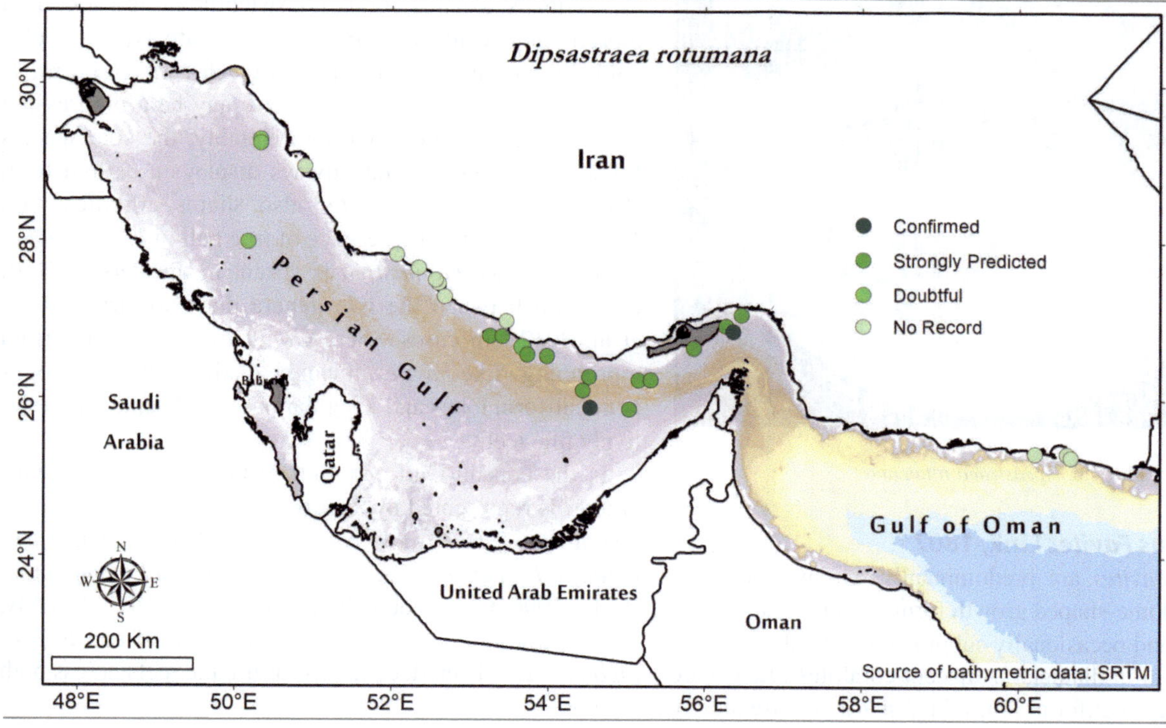

Classification: Phylum—Cnidaria, Class—Hexacorallia, Order—Scleractinia, Family—Merulinidae

Colony form: Colonies are typically massive and exhibit a rounded or semi-spherical shape (Plate 3.123). Their diameter ranges from 8 to 15 centimeters, occasionally reaching up to 30 centimeters.

Corallites: The polyp structure is organized into separate walls, with internal mesenteries (Plate 3.124). The polyp shapes are mostly semi-circular or irregular, often closely adjacent to each other, while the prominent walls remain distinct above the overall corallite structure. The minimum diameter of the corallite structure falls between 15 and 18 millimeters. The septa are multiple and arranged in 2 or 3 cycles. Notably, the primary septum is consistently prominent and curves above the corallite wall. Highly developed dentations create a rugged appearance, particularly near the corallite wall. Occasionally, secondary paliform lobes are present, adorned with coarse dentations. The columella is well-developed. Coenosteum often blistered.

Colour: These corals range from pale yellow to gray, with dark oral discs.

Plate 3.123 *Dipsastraea rotumana* from Larak Island

Distribution and abundance: This species is uncommon in the Persian Gulf and Gulf of Oman. In Iran is found mostly between 3 and 8 m around the offshore islands towards the Strait of Hormuz in the northern Persian Gulf.

Plate 3.124 Corallites in *Dipsastraea rotumana*

3.3.8.4 Genus *Favites* Link, 1807

Colonies of *Favites* are predominantly massive, adopting either flat or dome-shaped growth forms. Their corallites are monocentric and occasionally subplocoid, typically measuring 6 to 14 mm in diameter. Adjacent corallites often share common walls. Paliform lobes in *Favites* are frequently underdeveloped. Polyps extend only at night and possess a single circle of tapering tentacles, akin to those observed in *Dipsastraea*.

Favites exhibits similarities to *Dipsastraea* and *Goniastrea*. The primary distinction between *Favites* and *Dipsastraea* lies in the absence of distinct spacing between their calices. *Favites* corals consistently exhibit a cerioid morphology, while *Dipsastraea* corals are typically plocoid. In certain non-Gulf species (such as *F. flexuosa*, *F. complanata*, and *F. halicora*), there may be a progression toward a sub-cerioid condition. Notably, the sole species found in the Persian Gulf always displays a cerioid form (Riegl et al. 2012). *Favites* also shares affinities with *Goniastrea* due to the presence of true pali in *F. pentagona*. However, since *Goniastrea* is absent in the Persian Gulf, confusion between the two genera does not arise (Riegl et al. 2012). In cases where *Goniastrea* adopts a cerioid morphology like *Favites*, it can be distinguished by prominent paliform lobes and a regular pattern of septa with relatively fine teeth.

In the Persian Gulf and Gulf of Oman, the coral genus *Favites* is represented by ten distinct species: *F. flexuosa*, *F. pentagona*, *F. spinosa*, *F. chinensis*, *F. abdita*, *F. micropentagonus*, *F. acuticollis*, *F. ruselli*, *F. vasta*, and *F. complanata* (Coles 1996; Claereboudt 2006; Riegl et al. 2012). Notably, three species—*F. abdita*, *F. micropentagonus*, and *F. pentagona*—have been documented in Iranian reefs across both regions.

Favites abdita (Ellis & Solander, 1786)

Large star coral			
IUCN red list assessment		Status in Iran's reefs	
Status	Population trend	Abundance	Population trend
LC	Decreasing		Decreasing

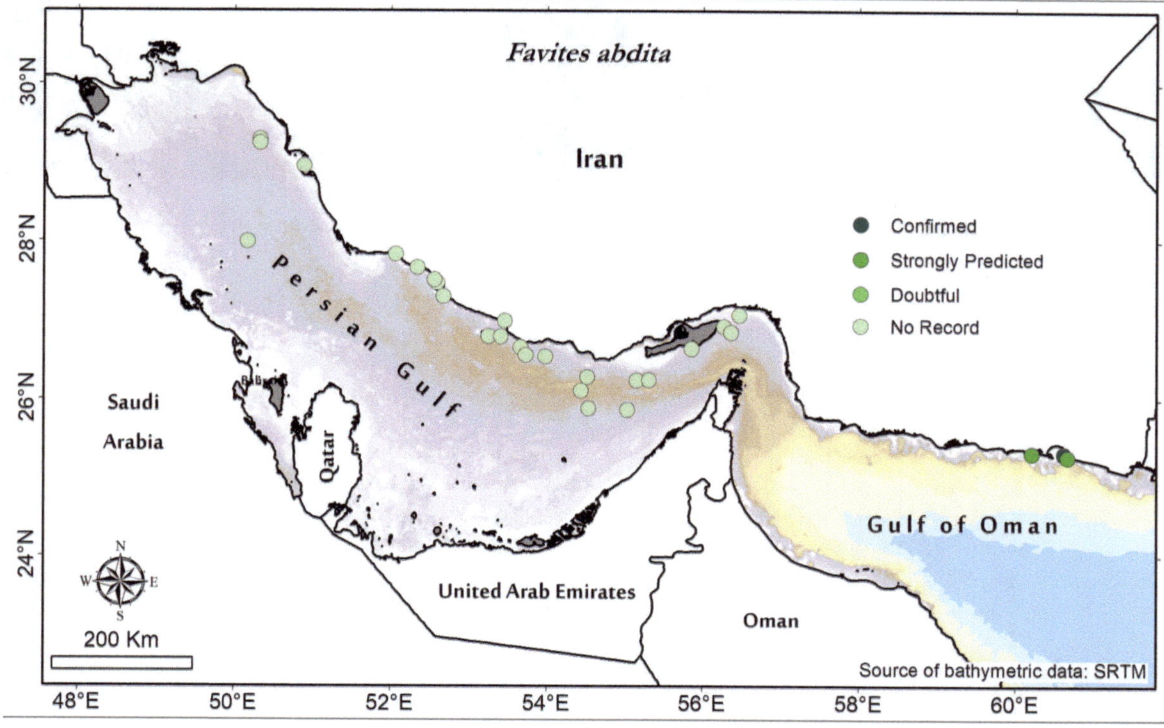

Classification: Phylum—Cnidaria, Class—Hexacorallia, Order—Scleractinia, Family—Merulinidae

Colony form: Colonies exhibit massive, rounded, and occasionally semi-spherical colonies (Plate 3.125). These colonies often develop regular, protruding hillocks, typically less than 30 centimeters in size, although occasionally reaching up to 50 centimeters or thereabouts.

Corallites: The polyp structure is organized into interconnected walls, with distinct and multi-sided corallites measuring up to 18 millimeters in diameter (Plate 3.126). The outer boundaries of the corallite structure are mostly semi-circular. Septa exhibit uniform thickness, bearing regular and well-developed trithecal teeth or even prominent ones. Paliform lobes are absent. Intriguingly, the intratentacular budding process is highly variable, resulting in significant size disparities within a single colony. It thrives in clear waters and commonly inhabits mid-slope depths ranging from 3 to 8 meters.

Colour: The color of *F. abdita* ranges from pale brown to brown.

Distribution and abundance: *Favites abdita* is the least abundant species in regions characterized by turbid waters or sediment deposition. This species is uncommoun across the Persian Gulf and Gulf of Oman.

Plate 3.125 *Favites abdita* from Chabahar Bay

Plate 3.126 Corallites in *Favites abdita*

Favites micropentagonus **(Veron, 2000)**

Golden war coral			
IUCN red list assessment		Status in Iran's reefs	
Status	Population trend	Abundance	Population trend
LC	Decreasing		Decreasing

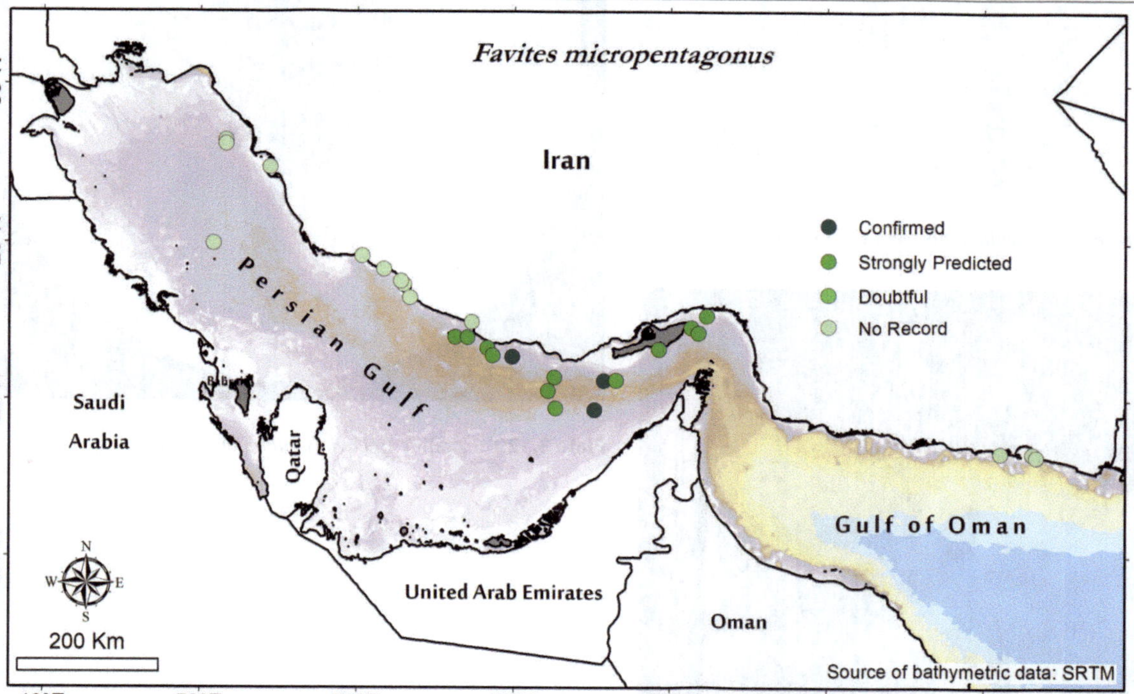

Classification: Phylum—Cnidaria, Class—Hexacorallia, Order—Scleractinia, Family—Merulinidae

Colony form: Colonies exhibit encrusting to sub-massive growth forms, occasionally nodular in appearance (Plate 3.127).

Corallites: The corallites are large, pentagonal in shape and measure 3–4 millimeters in diameter (Plates 3.128 and 3.129). Septa follow two alternating cycles and have irregular teeth. A well-developed paliform crown is present.

Colour: The overall coloration of colonies tends to be pale brown, sometimes with dark oral discs.

Distribution and abundance: This species is uncommoun across the Persian Gulf and Gulf of Oman.

Plate 3.127 *Favites micropentagonus* from Kish Island

Plate 3.128 Corallites in *Favites micropentagonus*

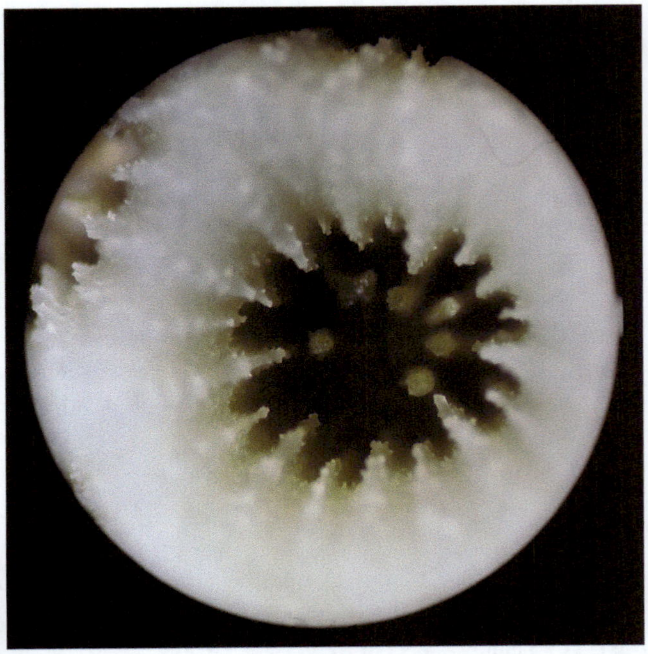

Plate 3.129 Corallite in *Favites micropentagonus*

Favites pentagona (Esper, 1790)

Lesser star coral			
IUCN red list assessment		Status in Iran's reefs	
Status	Population trend	Abundance	Population trend
LC	Decreasing		Decreasing

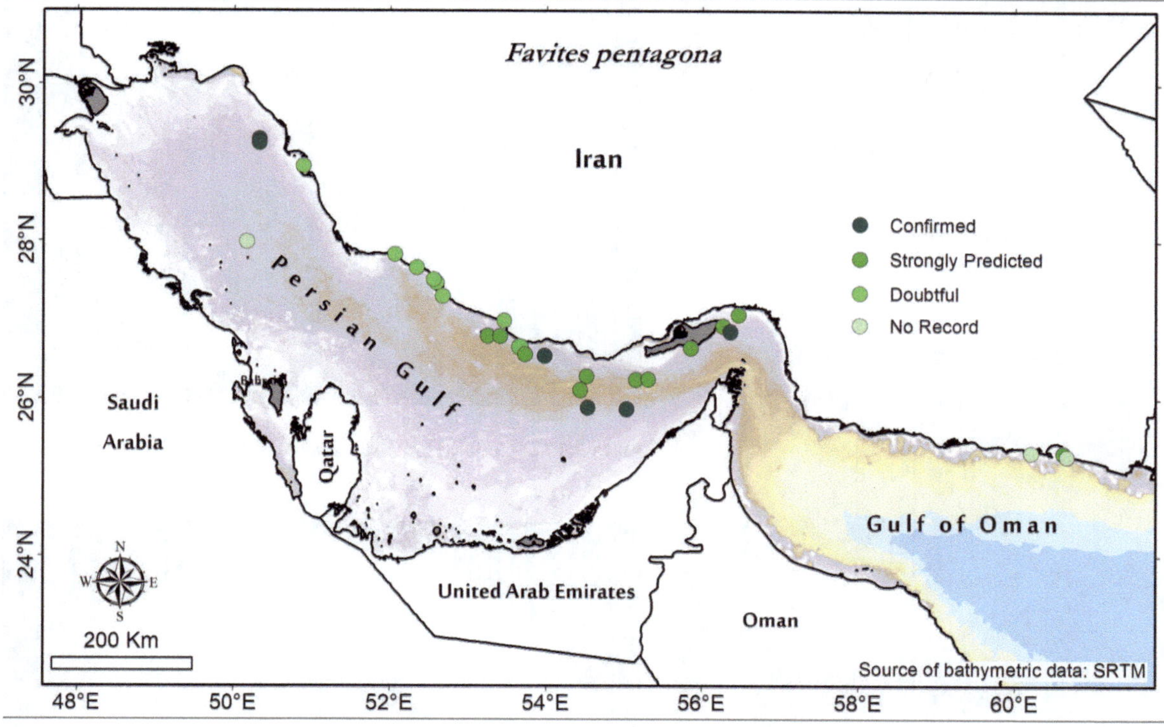

Classification: Phylum—Cnidaria, Class—Hexacorallia, Order—Scleractinia, Family—Merulinidae

Colony form: Colonies of this coral species exhibit submassive to encrusting growth forms, occasionally forming irregular columns (Plates 3.130 and 3.131).

Corallites: The corallites are characterized by thin walls and angular shapes, measuring less than 6 mm in diameter (Plates 3.132, 3.133, 3.134, and 3.135). Septa, arranged in two cycles, are dentate, with the first-order septa consistently reaching the columella. Second-order septa occasionally extend to the columella, while third-order septa are reduced to abortive structures. Notably, well-developed paliform lobes form a distinctive crown on the septa. These lobes often connect to the septum via trabecular linkages. The columella itself appears spongy, composed of twisted and fused trabeculae. The theca surrounding the corallites is thick. *Favites bestae* and *F. micropentagona* share similarities with this coral. The presence of the paliform crown further associates it with the *Goniastrea* genus.

Colour: The coral displays reddish or orange hues, often accompanied by lighter oral fields.

Distribution and abundance: Found in shallow reef environments. Although uncommon, this species occurs regularly throughout the entire Persian Gulf and Gulf of Oman.

3.3.8.5 Genus *Hydnophora* Fischer von Waldheim, 1807

The colonies of *Hydnophora* exhibit massive, encrusting, or arborescent growth forms. A distinguishing feature of this genus is the presence of hydnophores—conical mounds formed at the intersections of common walls between adjacent corallites. These hydnophores cover the colony surface, rendering *Hydnophora* immediately recognizable. Short tentacles surround the base of each hydnophore, with one tentacle positioned between each pair of septa.

Hydnophora has undergone taxonomic revisions based on macro-morphological traits such as wall and columella structure, as well as budding patterns. Initially grouped within the Faviidae family (alongside *Favites* and *Montastraea*) (Vaughan and Wells 1943; Veron et al. 1977), it was later reclassified into the Merulinidae (Veron 1986, 2000). However, molecular phylogenies reveal that *Hydnophora* is

Plate 3.130 *Favites pentagona* from Kish Island

Plate 3.131 *Favites pentagona* from Larak Island

more closely related to other Faviidae genera than to any included in the Merulinidae (Fukami et al. 2008).

The genus has been observed at Larak and Abu Musa, and it likely occurs on other islands closer to the Strait of Hormuz (Riegl et al. 2012). Additionally, *Hydnophora exesa* (Pallas, 1766) has been reported, albeit from a single specimen collected in Kuwait (Qaro Island) (Carpenter et al. 1997).

Three species have been documented in the Persian Gulf and Gulf of Oman, namely *H. exesa*, *H. pilosa*, and *H. microconos* (with the latter likely recorded from the Arabian Sea) (Coles 1996; Claereboudt 2006; Riegl et al. 2012). Notably, only one species, *H. pilosa*, has been formally described from the Iranian reefs.

Plate 3.132 Corallites in *Favites pentagona* from Kish Island

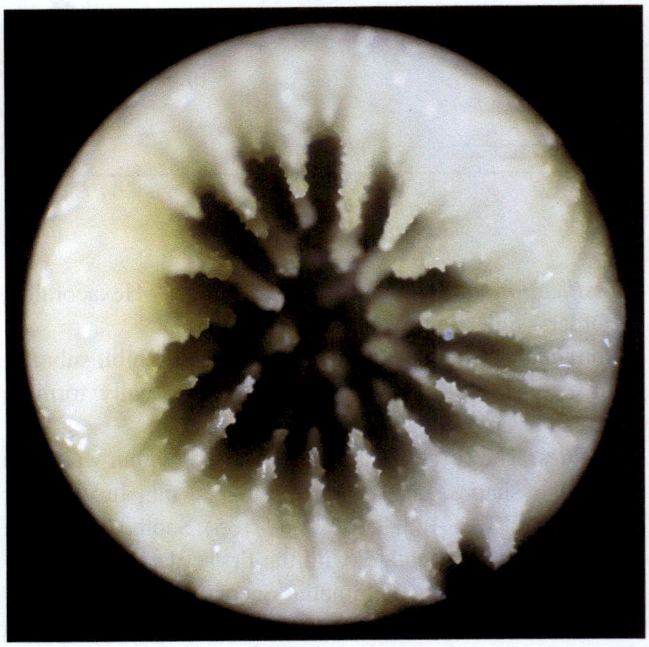

Plate 3.133 Corallite in *Favites pentagona* from Kish Island

Plate 3.134 Corallites in *Favites pentagona* from Larak Island

Plate 3.135 Corallite in *Favites pentagona* from Larak Island

Hydnophora pilosa **(Veron, 1985)**

Spine coral			
IUCN red list assessment		Status in Iran's reefs	
Status	Population trend	Abundance	Population trend
LC	Decreasing		Decreasing

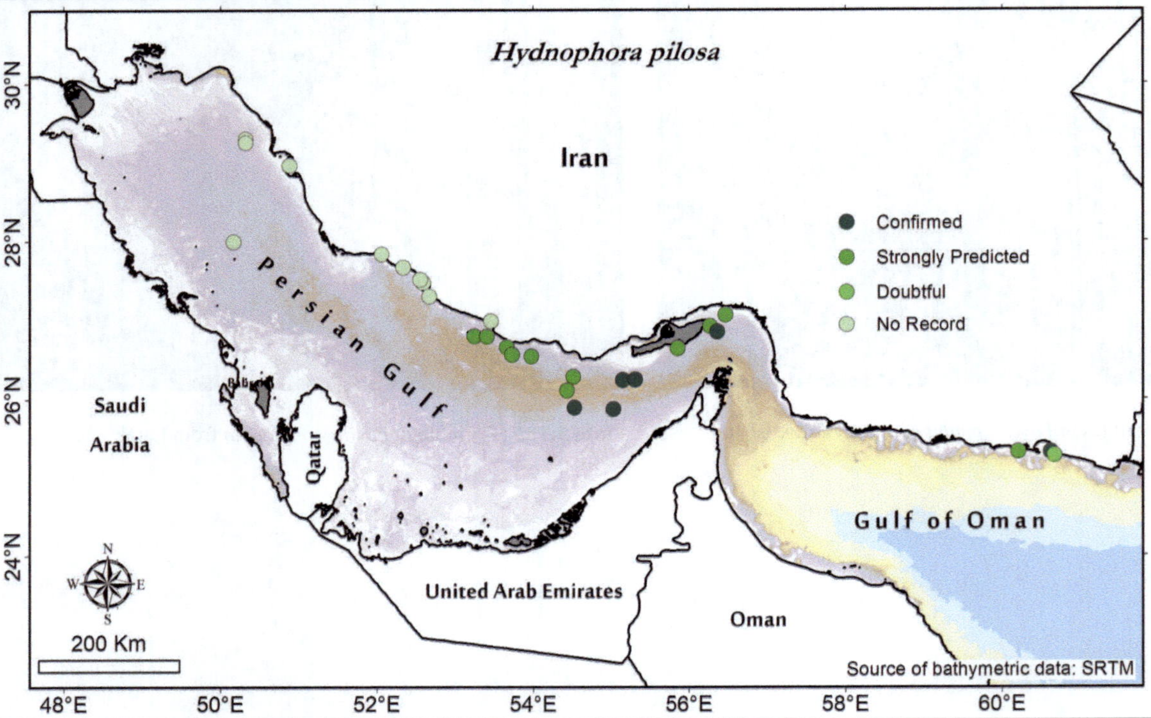

Classification: Phylum—Cnidaria, Class—Hexacorallia, Order—Scleractinia, Family—Merulinidae

Colony form: The corallum exhibits an encrusting, laminar or sub-massive growth form, characterized by knobbly protuberances (Plate 3.136). These features resemble those of *Hydnophora exesa*, although the latter species may have more developed branches and more exsert monticules.

Corallites: Within the corallites, which are hydnophoroid, well-developed monticules of varying size and orientation can form single polyps or series (Plate 3.137). The number of septa reaching the tips of these monticules varies significantly. Typically, the monticule axis aligns perpendicularly to the surface of the corallum. The septa exhibit a uniform size class, usually representing the first complete cycle. Smaller monticules typically have six septa, while longer, fused ones tend to exhibit multiples of six. Secondary intermediate septa are rarely observed. These septa maintain uniform thickness throughout their length and bear well-developed teeth and denticular ornamentations. The columella is usually present as a series of twisted trabeculae in the valleys between monticules.

Colour: The polyps exhibit dark brown or blue-grey coloration, while their tentacles feature white tips.

Distribution and abundance: The distribution of this coral includes Larak Island and is likely to be found in other islands near the Strait of Hormuz. If the Qaro specimen (previously assigned to *H. exesa* by Carpenter et al. 1997) indeed represents the same species, then its occurrence extends throughout the Persian Gulf.

3.3.8.6 Genus *Platygyra* Ehrenberg, 1834

Platygyra colonies exhibit massive growth forms, which can be either flat or dome-shaped. The corallites are rarely cerioid but commonly meandroid. Colony formation occurs through linear intramural and polystomodeal budding. Thecae are narrow, and those of adjacent valleys fuse together. Paliform lobes are absent, and the columellae seldom form distinct centers, instead appearing as a continuous tangle of species. Polyps extend only during the nighttime.

Plate 3.136 *Hydnophora pilosa* from Larak Island

Plate 3.137 Corallites in *Hydnophora pilosa*

Platygyra shares similarities with *Goniastrea* and *Leptoria*. *Leptoria*, more meandroid than *Platygyra*, features distinctive wall-like columellae and uniformly spaced septa of equal size. Across environmental gradients, all *Platygyra* species exhibit similar skeletal modifications. However, distinguishing between certain species, such as *P. daedalea* and *P. lamellina*, can be challenging, especially when they co-occur. Recent genetic and morphometric analyses suggest that perhaps only one valid species, namely *P. daedalea*, exists (Miller 1992; Miller and Babcock 1997; Miller and Benzie 1997). While some records list *P. crosslandi* and *P. sinensis* occurring in the Persian Gulf (Sheppard and Sheppard 1991; Riegl 1999; Veron 2000), insufficient justification exists to verify these claims. Morphological variability allows for maintaining a separation between *P. daedalea* and *P. lamellina*, as proposed by Veron (2000). In the field, distinguishing between the two species based on skeletal characters is often impractical due to tissue obscuration, with the shorter series in *P. daedalea* serving as a distinguishing feature.

Platygyra daedalea **(Ellis & Solander, 1786)**

Brain coral			
IUCN red list assessment		Status in Iran's reefs	
Status	Population trend	Abundance	Population trend
LC	Decreasing		Stable

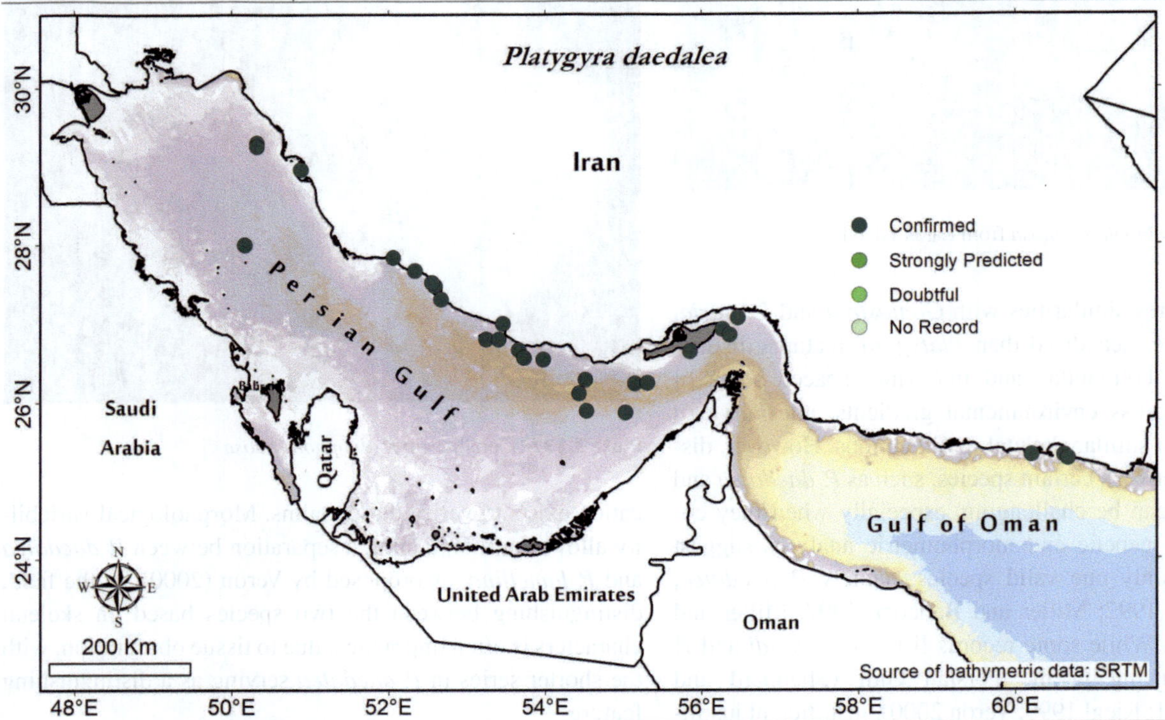

Classification: Phylum—Cnidaria, Class—Hexacorallia, Order—Scleractinia, Family—Merulinidae

Colony form: This species exhibits a massive corallum structure, ranging from flat to hemispherical or occasionally encrusting formations (Plates 3.138, 3.139, and 3.140).

Corallites: The corallites are distinctly meandroid, characterized by long and deep valleys with marked centers that are only present at the termini of valleys or intermittently in short sequences (Plates 3.141, 3.142, and 3.143). Valley walls are generally narrow and frequently perforated. There is variability in valley length, with most specimens from the Persian Gulf corresponding to the variability previously attributed to *P. lamellina*. The septa are arranged in one or two orders, with most displaying a uniform exsertion and fusion above the theca, thereby forming continuous septal connections between adjacent valleys. At the base of the septa and above the theca, a fine trabecular linkage is observed, contributing to the well-defined dentations; however, paliform lobes are typically absent. The columella is robustly developed, exhibiting a trabecular and spongy texture, although its development may vary within individual colonies.

Colour: The coloration of this species is variable and often vibrant, with common patterns including brown walls and green valleys; darker hues are observed on shaded sides of colonies.

Distribution and abundance: This species is one of the most prevalent corals, occupying diverse habitats throughout the Persian Gulf and Gulf of Oman.

3.3.9 Family Fungiidae Dana, 1846

The Fungiidae family, known as mushroom or plate corals, comprises 16 extant genera (Hoeksema and Cairns 2024a) ranging from solitary to colonial species, predominantly hermatypic. Colonial forms have evolved from solitary ancestors, retaining septo-costal structures that extend radially as septa on the upper surface and costae underneath. Anthocaulus marks the initial attached phase, transitioning to the free-living Anthocyathus stage. Despite occasional bleached or deceased appearances, most genera feature a central polyp that emerges nocturnally to feed. While most species remain fully detached from the substrate upon reaching adulthood, some exhibit immobility and colonial growth

Plate 3.138 *Platygyra daedalea* from Larak Island

Plate 3.139 *Platygyra daedalea* from Kharg Island

Plate 3.140 *Platygyra daedalea* from Kish Island

Plate 3.141 Corallites in *Platygyra daedalea* from Larak Island

Plate 3.142 Corallites in *Platygyra daedalea* from Kharg Island

patterns. Certain genera such as *Cycloseris* and *Fungia* are solitary organisms; *Polyphyllia* consists of a single organism with multiple mouths; *Ctenactis* and *Herpolitha* could be perceived as either solitary organisms with multiple mouths or as colonies of individual polyps each possessing its own mouth.

Plate 3.143 Corallites in *Platygyra daedalea* from Kish Island

Out of 16 genera identified globally, *Cycloseris* and *Fungia* are reported from the Persian Gulf (Riegl et al. 2012) and Gulf of Oman (Claereboudt 2006) as lifelong solitaries. Although 51 species are recognized worldwide, only one species is observed within Iranian reefs in these regions.

3.3.9.1 Genus *Cycloseris* Milne Edwards & Haime, 1849

Species of the genus *Cycloseris* are typically solitary and exhibit a flat, circular, or slightly oval morphology with a central oral opening. The septa are characterized by delicate dentition, while the costae are finely structured and lack pits on their undersurface. Polyp extension predominantly occurs nocturnally, with slender tentacles adorning the upper surface of the disc. *Cycloseris* bears resemblance to the genera *Fungia* and *Diaseris*; however, it can be distinguished by its comparatively smaller size. *Fungia* species tend to achieve larger dimensions, may present an elongated form, and possess septa with more pronounced dentition as well as costae formed by aligned spines. In contrast, *Diaseris* is identified by its wedge-shaped segments that contribute to an overall irregular configuration, alongside septa that are robust with blunt dentition.

Two species have been documented in the Persian Gulf, namely *C. explanulata* (present study), and *C. curvata* (Riegl et al. 2012) both from Iranian offshore reefs towards the Strait of Hormuz.

Cycloseris explanulata **(Van der Horst, 1922)**

Flat coral			
IUCN red list assessment		Status in Iran's reefs	
Status	Population trend	Abundance	Population trend
LC	Decreasing		Decreasing

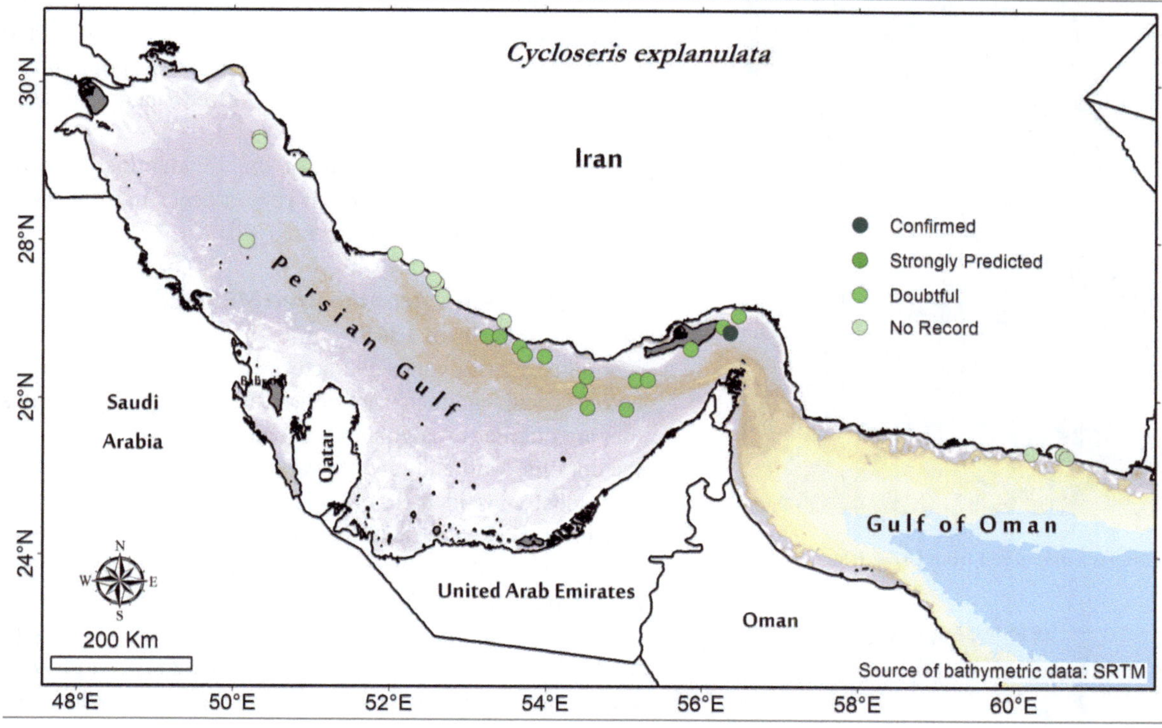

Classification: Phylum—Cnidaria, Class—Hexacorallia, Order—Scleractinia, Family—Fungiidae

Colony form: The colonies are characterized by their thin plate-like structure with large, evenly distributed corallites and are enveloped in thin coverings ranging from 5 to 10 centimeters in diameter (Plate 3.144).

Corallites: Polyp structures with shared walls measure between 2 to 3 millimeters in diameter (Plate 3.145). Septa transition from one center to another, merging and exhibiting significant ornamentation and alternation. The thicker septa are more pronounced and often possess petaloid tentacles as seen in many species within the mushroom coral family. Petaloid septa protrude above the colony's surface. The columella consists of 8 to 12 rounded protrusions, with tentacles typically extending during daylight hours.

Colour: The coloration includes variegated green and brown hues, sometimes brick red or terracotta, with tentacle tips usually appearing white.

Distribution and abundance: This species is not readily noticeable due to its coloration and tiny colony size. Within the Iranian reefs, it is considered relatively rare and predominantly occupies the shallow reefs surrounding the offshore islands and reef banks towards the Strait of Hormuz in the northern Persian Gulf.

3.3.10 Family Lobophylliidae Dai & Horng, 2009

The family Lobophylliidae represents a group of large polyp stony corals. Its establishment in 2009 followed a comprehensive revision of the so-called "robust" coral families, including Faviidae, Merulinidae, Mussidae, and Pectiniidae. This revision revealed that these families were polyphyletic, meaning their similarities were due to convergent evolution rather than shared ancestry. The Indo-Pacific species traditionally classified within Mussidae contributed to the formation of the new family Lobophylliiae. Additionally, some species previously grouped under Pectiniidae were incorporated into Lobophylliiae, while the remaining Pectiniidae species merged into Merulinidae (Budd et al. 2012). The type genus for Lobophylliiae is *Lobophyllia*.

Traditionally, the "robust" coral families (Faviidae, Merulinidae, Mussidae, and Pectiniidae) were distinguished based on morphological characteristics. However, recent

Plate 3.144 *Cycloseris explanulata* from Larak Island

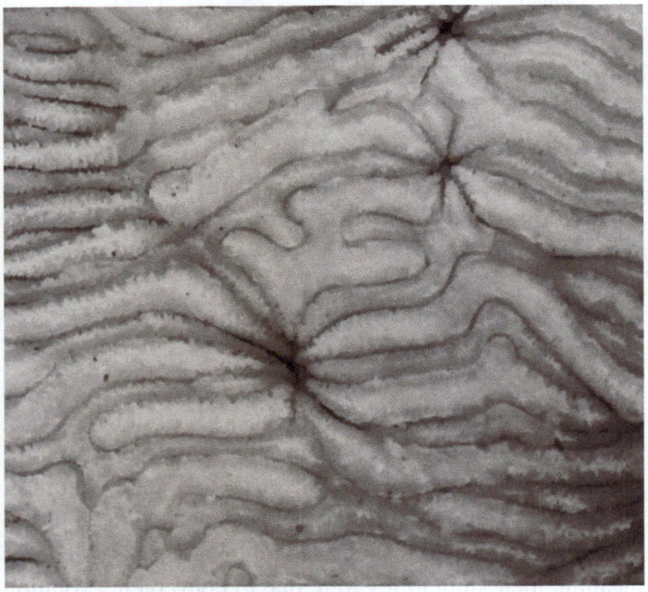

Plate 3.145 Corallites in *Cycloseris explanulata*

molecular analyses revealed their polyphyletic nature, indicating that their similarities arose independently. Notably, genera like *Favia* and *Scolymia* were also found to be polyphyletic, with Atlantic *faviids* and *scolymids* more closely related to each other than to their Indo-Pacific counterparts. In 2012, a revised classification placed Pacific Mussidae species within the newly established family Lobophylliiae. The taxon Mussidae was retained for Atlantic species (Budd et al. 2012). Genera such as *Echinomorpha*, *Echinophyllia*, and *Oxypora* were transferred from Pectiniidae to Lobophylliiae, along with the genus *Moseleya* (Goffredo and Dubinsky 2016).

Regarding distribution, out of the 13 identified genera, four species from four different genera occur in the Persian Gulf and Gulf of Oman (Riegl et al. 2012).

3.3.10.1 Genus *Acanthastrea* Milne Edwards & Haime, 1848

Acanthastrea is a genus characterized by massive, often flat colonies with monocentric, cerioid or subplocoid corallites that are circular or angular. The septo-costae are robust near the corallite wall and taper towards the columella, featuring prominent mussid teeth. Polyps are nocturnal and fleshy, obscuring skeletal structures when extended, which aids in distinguishing *Acanthastrea* underwater. *Acanthastrea echinata* is particularly notable for its resemblance to *Favites* in skeletal structure but can be differentiated by its unique septal dentation and columella structure. *Acanthastrea maxima* stands out as a distinct species within the genus. Within the Persian Gulf and Gulf of Oman, the genus *Acanthastrea* is represented by four distinct species: *A. echinata*, *A. hillae*, *A. maxima* and *A. hemprichii* (Coles 1996; Carpenter et al. 1997; Riegl et al. 2012; Vajed Samiei et al. 2013). Among these, solely *A. echinata* has been delineated within this volume.

Acanthastrea echinata **(Dana, 1846)**

Starry cup coral			
IUCN red list assessment		Status in Iran's reefs	
Status	Population trend	Abundance	Population trend
LC	Decreasing		Decreasing

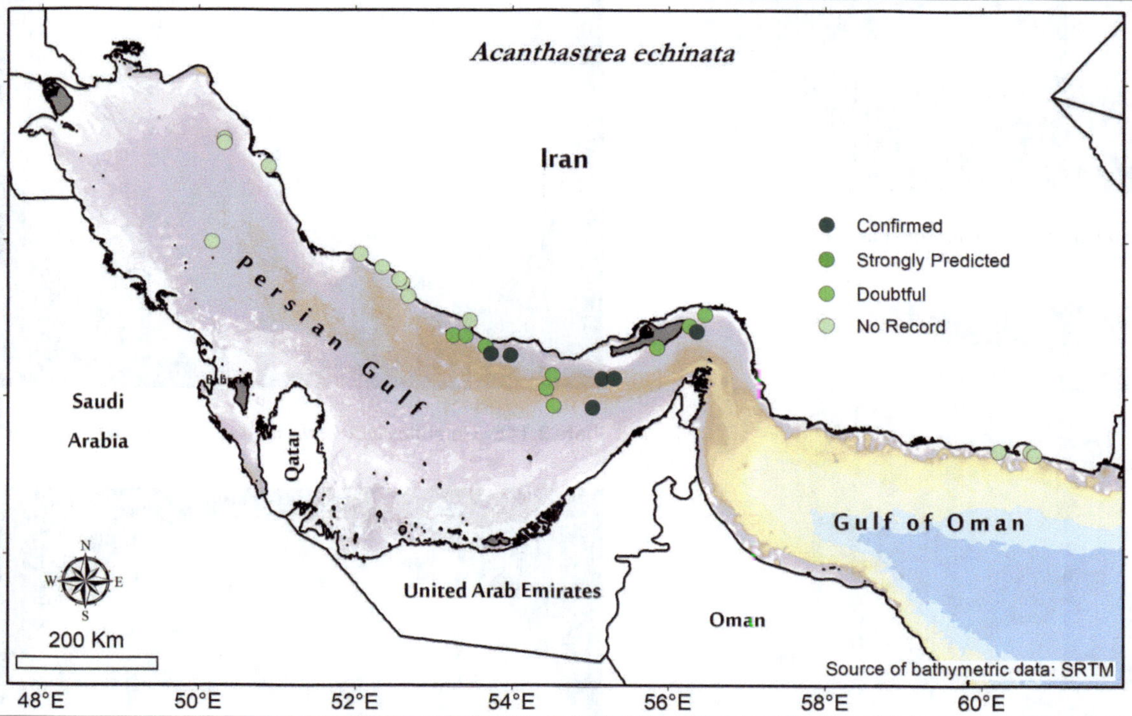

Classification: Phylum—Cnidaria, Class—Hexacorallia, Order—Scleractinia, Family—Lobophylliidae

Colony form: This species exhibits a corallum that is either encrusting or massive in form (Plates 3.146 and 3.147).

Corallites: The corallites are cerioid, ranging from round to oval, with a diameter typically exceeding 10 mm, characterized by intratentacular budding (Plates 3.148 and 3.149). Septa are singularly ordered but vary in number with corallite size; most septa extend to the columella, with some fusing at their inner margins. These septa are evenly spaced and feature long, irregular, echinulate dentations on their upper edges. The septocostae are pronounced, interlinking with those of adjacent corallites. The columella is prominent, composed of twisted septal dentations.

Colour: The coloration spans from brown to green, often mottled.

Distribution and abundance: In terms of habitat preferences, *A. echinata* is consistently found throughout a wide range of depths and environmental conditions, with a distribution depth reaching up to 20 meters, and demonstrates an adaptability to waters with high turbidity. Although it is extensively distributed in the Persian Gulf and Gulf of Oman, its occurrence is uncommon.

3.3.10.2 Genus *Oxypora* Saville Kent, 1871

Oxypora, a genus within the large polyp stony coral family, forms colonies that are predominantly foliaceous with very thin laminae. The calices of *Oxypora* are typically round or oval, irregular, and shallow, with a non-prominent inclination on the corallum surface. The septa are sparse and the columellae are underdeveloped, with the coenosteum exhibiting pits at the juncture of new septo-costae insertions. Distinguishing *Oxypora* from *Echinophyllia* can be challenging, particularly when colonies thicken; it may also be mistaken for *Echinopota* and *Mycedium*. Indigenous to the Indo-Pacific region, of the five species recognized globally (Hoeksema and Cairns 2024b), only *O. lacera* is documented in the Persian Gulf and Gulf of Oman (Coles 1996; Claereboudt 2006; Riegl et al. 2012).

Plate 3.146 *Acanthastrea echinata* from Larak Island

Plate 3.147 *Acanthastrea echinata* from Kish Island

Plate 3.148 Corallites in *Acanthastrea echinata* from Larak sland

Plate 3.149 Corallites in *Acanthastrea echinata* from Kish Island

Oxypora lacera (Verrill, 1864)

Porous lettuce coral			
IUCN red list assessment		Status in Iran's reefs	
Status	Population trend	Abundance	Population trend
LC	Decreasing		Unknown

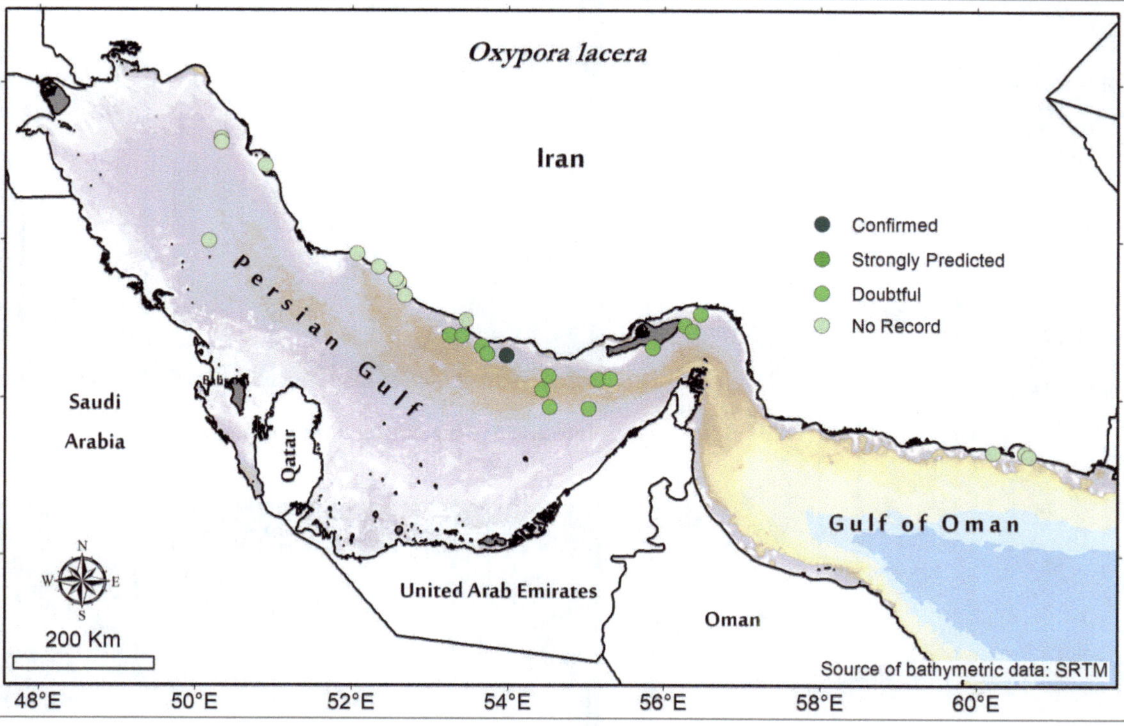

Classification: Phylum—Cnidaria, Class—Hexacorallia, Order—Scleractinia, Family—Lobophylliidae

Colony form: The colonies are either single-layered thin plates or irregularly shaped with convoluted edges or vertical expansions, reaching a maximum diameter of 40 centimeters (Plates 3.150, 3.151, and 3.152). Colonies may exhibit significant thickening in turbulent conditions.

Corallites: The corallites range from fine and delicate on thin laminae to markedly thickened on the more substantial parts of the colony (Plates 3.153 and 3.154). Polyp structures are superficial, lacking true walls, with an average diameter ranging between 4 to 8 millimeters. Occasionally, the initial polyp structure is visible at the colony's center, distinctly larger than others, with a diameter of 12 to 15 millimeters. Polyp structures are well-separated from each other. The septal ornamentation consists of a few well-developed teeth with pointed ends. The costae are invariably toothed, with intercostal slits typically absent from the colony's central regions. *Oxypora lacera* generally possesses fewer than 12 primary septa and is distinguished by its less inclined corallites near the periphery and radial symmetry near the colony center. The coenosteum is elaborately decorated with flattened ribs bearing sharp but infrequent dentations. Around the colony's perimeter, there are coenosteum holes resembling slits, generally found in pairs near where the rib terminates. These become obscured and appear as slightly elongated air vesicles in the colony's central part. In terms of habitat and abundance, *O. lacera* is a shade-loving species often found on mid-slopes or deeper areas, particularly beneath overhangs, as well as in pits and crevices.

Colour: The species is characterized by a color palette that usually includes pale brown or greenish hues, with variations that may feature green, white, or red oral discs.

Distribution and abundance: It is not abundant and is considered relatively uncommon in the Persian Gulf and Gulf of Oman. This species predominantly inhabits depths exceeding 10 meters in sheltered areas that are characterized by minimal sedimentation and limited currents.

Plate 3.150 *Oxypora lacera* from Kish Island

Plate 3.151 *Oxypora lacera* from Kish Island

Plate 3.152 Bleached *Oxypora lacera* from Kish Island

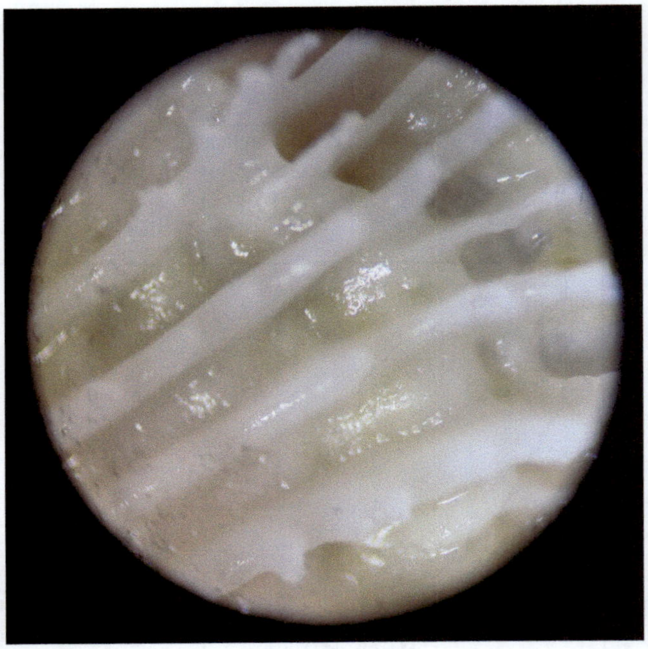

Plate 3.153 Plate surface in *Oxypora lacera*

Plate 3.154 Corallites in *Oxypora lacera*

3.3.10.3 Genus *Sclerophyllia* Klunzinger, 1879

The genus *Sclerophyllia* comprises stony corals characterized by their solitary and predominantly circular morphology with calices reaching up to 15 mm in diameter. Possesses thick walls with septa that lack a specific order. The septa are robust near the calice rim, tapering and exhibiting strong dentation. Columella is large, spongy, and deeply situated within the calice. In terms of pigmentation: exhibits a color range from cream to pale brown. These corals typically inhabit reef slopes. This genus is represented by one species "*S. maxima*" in the Persian Gulf and Gulf of Oman.

Sclerophyllia maxima (Sheppard & Salm, 1988)

Fleshy artichoke coral			
IUCN red list assessment		Status in Iran's reefs	
Status	Population trend	Abundance	Population trend
VU	Decreasing		Decreasing

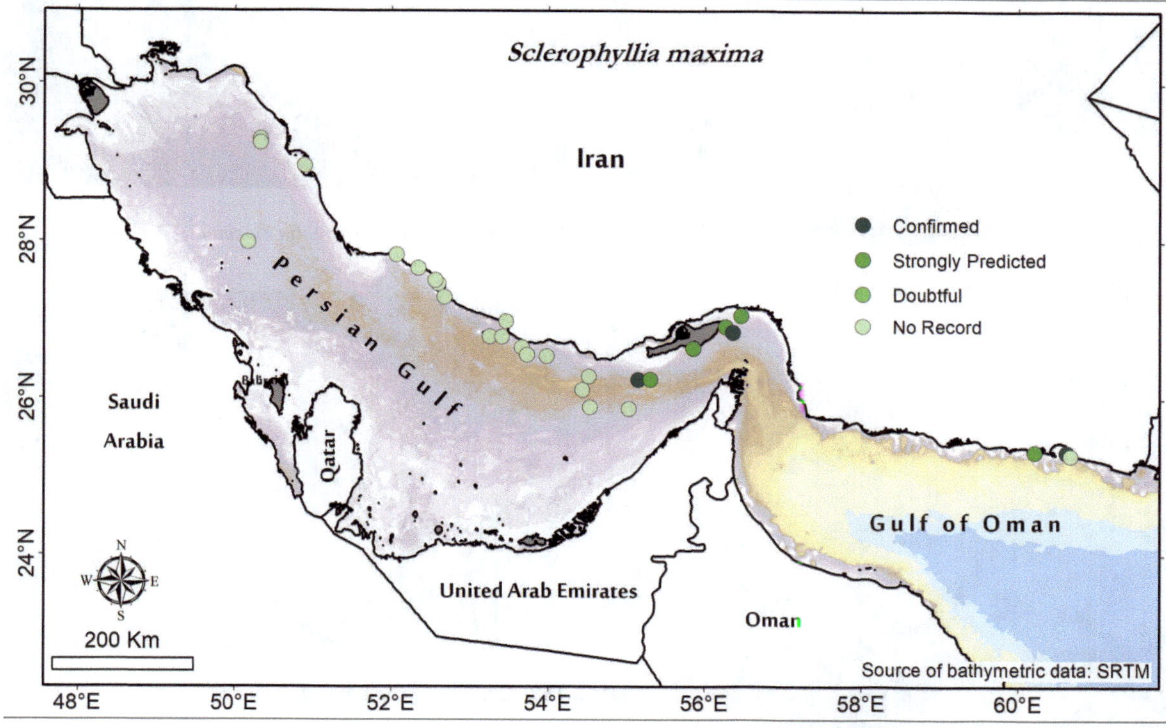

Classification: Phylum—Cnidaria, Class—Hexacorallia, Order—Scleractinia, Family—Lobophylliidae

Colony form: The corallum of this species ranges from submassive to massive (Plates 3.155 and 3.156).

Corallites: The corallites are cerioid, round to oval, with diameters typically around 10 mm, often larger (Plates 3.157 and 3.158). Intratentacular budding occurs at the periphery of the corallites, giving an almost extratentacular appearance (Sheppard and Salm 1988). Corallites are always monocentric, with diameters up to 50 mm and depths up to 10 mm, making them the largest corallites of any coral in the Persian Gulf and Gulf of Oman. The septa are organized into at least three orders: the first order reaches the columella, the second order extends to about half the radius, and the third order consists of thin ridges or abortive structures. Large calices can have up to 80 septa, with upper septal margins featuring large dentations and weak exsertions (1–3 mm). The columella is well-developed, comprising twisted septal dentations. The thecae are also well-developed, with thick, vesicular endotheca and a ridge running along the top.

Colour: The color ranges from brown to green.

Distribution and abundance: This species is uncommon but widely distributed throughout the Persian Gulf, recorded in Oman (type locality), Iran, and Kuwait. Within the Iranian reefs, it is considered relatively rare and predominantly occupies the shallow reefs surrounding the offshore islands and reef banks towards the Strait of Hormuz in the northern Persian Gulf. It is also recorded in Chabahar Bay in the Gulf of Oman.

3.3.10.4 Genus *Echinophyllia* Klunzinger, 1879

The genus *Echinophyllia* comprises large polyp stony corals. These colonial corals typically exhibit a foliaceous growth form, characterized by very thin, leaf-like structures. Colonies are either encrusting or laminar, with calices that are round or oval, immersed to tubular, and not strongly inclined on the colony surface. The septa are usually numer-

Plate 3.155 *Sclerophyllia maxima* from Chabahar Bay

Plate 3.156 *Sclerophyllia maxima* from Chabahar Bay

Plate 3.157 Corallites in *Sclerophyllia maxima*

Plate 3.158 Corallites in *Sclerophyllia maxima*

ous, columellae are well developed, and the coenosteum is pitted at the commencement of new septo-costae. These corals can be found in various shades of brown, green, or red, often with contrasting colors on the oral discs of the polyps.

Native to the Indo-Pacific region, *Echinophyllia* corals are commonly found in reef aquariums. They inhabit a range of depths, from shallow reef crests to deeper, more sheltered fore reef habitats. This genus is represented by the species *E. aspera* throughout the Persian Gulf and Gulf of Oman (Coles 1996; Claereboudt 2006).

Echinophyllia aspera **(Ellis & Solander, 1786)**

Flat lettuce coral			
IUCN red list assessment		Status in Iran's reefs	
Status	Population trend	Abundance	Population trend
LC	Decreasing		Decreasing

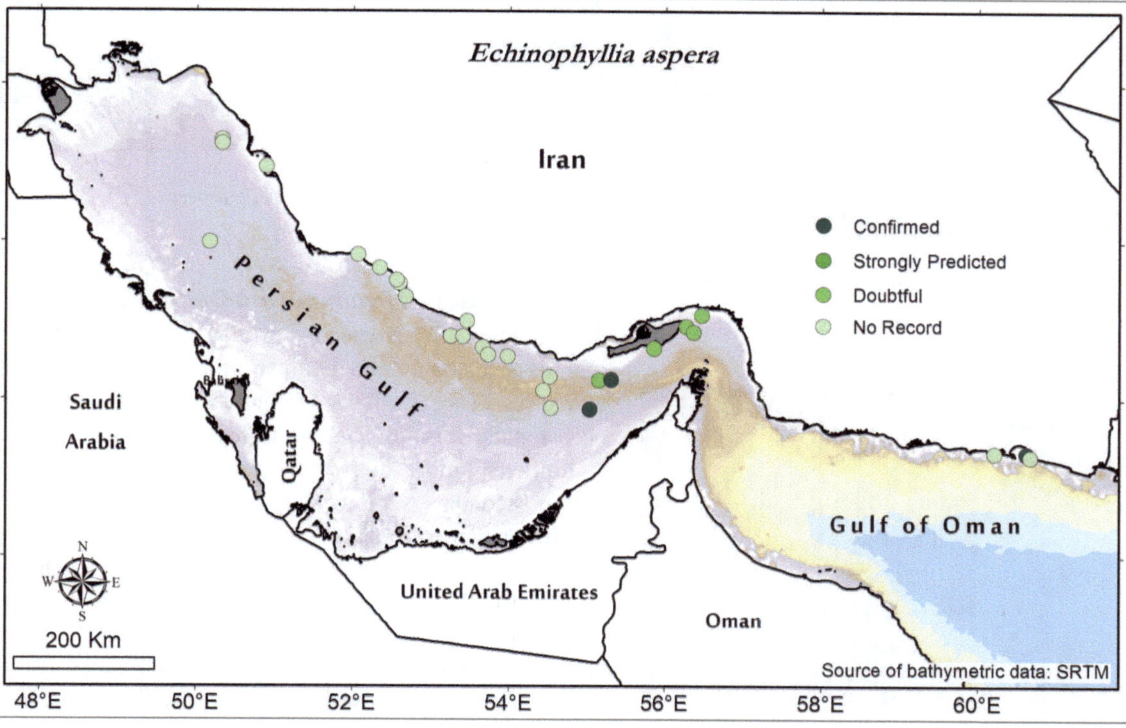

Classification: Phylum—Cnidaria, Class—Hexacorallia, Order—Scleractinia, Family—Lobophylliidae

Colony form: The corallum ranges from encrusting to partially foliose, with the central part often knobby and featuring protruding corallites (Plate 3.159). Colonies can reach up to 50 cm in diameter, sometimes partially covering the substrate.

Corallites: Small-sized specimens often have a central knobby structure with distinct corallites (Plates 3.160 and 3.161). The average diameter of corallites ranges from 5 to 8 mm, sometimes slightly inclined towards the edge of the corallum, and are well-separated by a wide coenosteum. The distance between corallites ranges from 2 to over 50 mm. The septa are present in one or two orders, varying in number from 14 to 20, and are typically wedge-shaped, thicker near the corallite wall, and tapering towards the columella. They are uniformly prominent, extending up to 5 mm above the poorly developed wall, with well-developed dentate margins. The columella is present and consists of a few small trabeculae. The costae are well-developed, extending across the coenosteum, and are often closely spaced and perpendicular to the corallum edge in smaller specimens, and towards the outer edges in larger specimens, with numerous fine dentations.

Colour: The coloration is generally grey or blue-grey, occasionally grey-green or brown.

Distribution and abundance: This species has not been observed in the southern part of the Persian Gulf but can be found in the northern part, from the Strait of Hormuz to Kharg and Khargoo islands. This species has been documented on the reefs of Oman in the Gulf of Oman (Coles 1996).

3.3.11 Family Psammocoridae Chevalier and L. Beauvais, 1987

The Psammocoridae family, described by Chevalier and Beauvais in 1987, was created to accommodate the genus *Psammocora* based on its distinctive corallite morphology. Despite initial disregard by some researchers (Veron and Pichon 1976; Veron 2000), subsequent molecular phylogenies and re-examinations of polyp and corallite morphology

Plate 3.159 Bleached colony of *Echinophyllia aspera* from Kharg Island

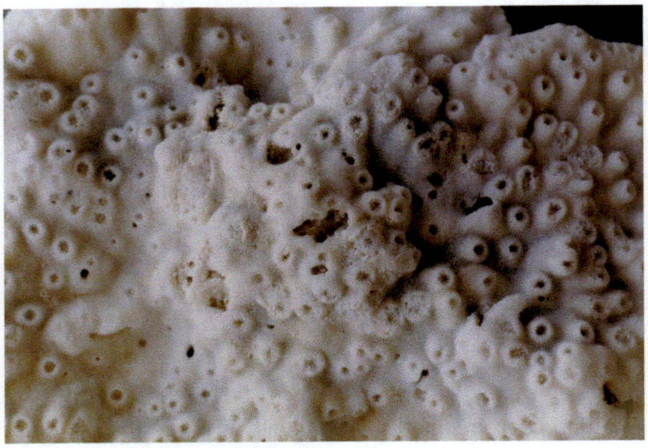

Plate 3.160 Corallites of *Echinophyllia aspera*

have confirmed the distinctiveness of the Psammocoridae and their close relationships with certain genera previously classified under the Siderastreidae (Benzoni et al. 2007, 2010; Fukami et al. 2008). Members of this monotypic family, commonly known as cat's paw coral, are characterized by thick plates with low, irregular ridges and small, shallow corallites that are irregularly distributed. This family includes a single genus, *Psammocora*, which is characterized by a unique pattern of septal fusion and star-like valleys.

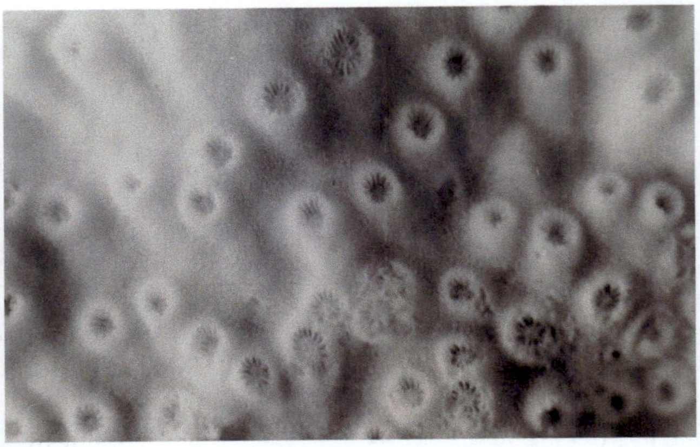

Plate 3.161 Corallites of *Echinophyllia aspera*

3.3.11.1 Genus *Psammocora* Dana, 1846

The colonies of *Psammocora* exhibit various forms, including massive, columnar, laminar, foliaceous, or encrusting structures. The corallites are notably small and shallow, occasionally forming shallow valleys, with indistinct walls. A limited number of primary septo-costae are embedded within secondary septo-costae, creating distinctive species-specific patterns. Polyps typically extend only at night. *Psammocora* is often confused with *Coscinaraea*, which can be initially distinguished by its larger corallites and calices. Polyps in *Psammocora* are characterized by concentric crowns of tentacles, with only the innermost tentacles directly connected to the polyp gastrovascular cavity. Within the calice, entosepta (bearing a tentacle) alternate with exosepta (devoid of tentacles) and are interconnected by synapticulae. The detailed pattern of entosepta and exosepta furcation and fusion is elaborated in Benzoni et al. (2007). Entosepta and enclosed entosepta can be thicker than exosepta, attaining a characteristic "petaloid" shape, which is distinctive of this genus. Five species of *Psammocora* are recorded in the Persian Gulf and Gulf of Oman, including *P. contigua*, *P. haimeana*, *P. superficialis*, *P. stellata*, and *P. albopicta* (Coles 1996; Riegl et al. 2012). *Psammocora albopicta*, recently described and previously confused with *P. superficialis*, and *P. stellata*, whose small colonies may resemble *P. profundacella* before developing the typical branching morphology, are notable (Riegl et al. 2012). In Iranian reefs across the Persian Gulf and Gulf of Oman, three species are present: *P. albopicta*, *P. contigua*, and *P. columna*.

Psammocora albopicta (Benzoni, 2006)

IUCN red list assessment		Status in Iran's reefs	
Status	Population trend	Abundance	Population trend
LC	Decreasing		Unknown

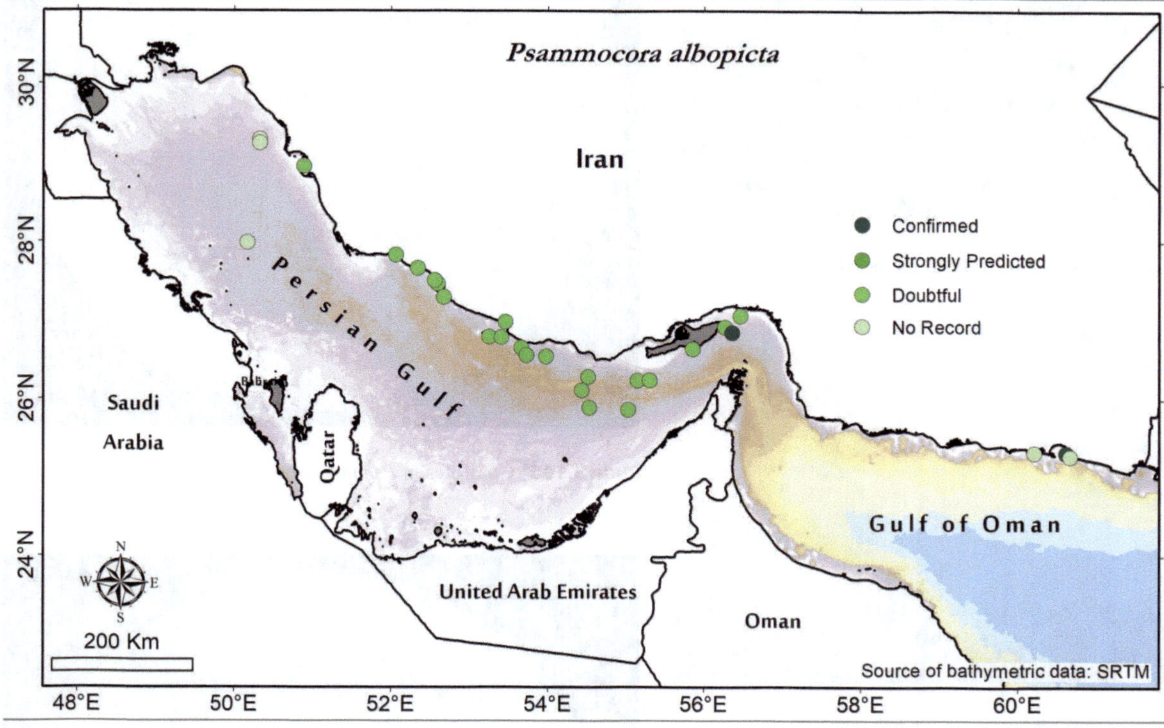

Classification: Phylum—Cnidaria, Class—Hexacorallia, Order—Scleractinia, Family—Psammocoridae

Colony form: The corallum is generally encrusting to submassive, conforming to the shape of the substrate (Plates 3.162, 3.163, and 3.164). The colony diameter ranges from 5 cm to 60 cm. The colony surface is smooth or undulating, with knobs or nodules scattered across it.

Corallites: The corallites are very small, with a diameter of about 1 mm, and are barely visible underwater (Plates 3.165 and 3.166). The distances between corallites are equal, and they can form short series at times, separated by up to two rows of enclosed septa. Shallow depressions, delineated by rounded collines or ridges, may develop in parts of the corallum, though these ridges are never as acute as in *P. stellata*. Seven to nine septa reach the fossa, with three to four being petaloid or ribbon-like. The wall is rarely visible unless it protrudes from the colony surface to form a circular series of hills surrounding the corallites or a short series of corallites. The columella is present, consisting of a single process that is round, oval, or nearly cross-shaped in section. The tentacles usually extend during the day, and the columella consists of single, round projections.

Colour: The color ranges from light to dark brown, often with characteristic light-colored areas, giving the species its name "white-spotted." Due to its pale color and very small polyp skeleton, *P. albopicta* is considered somewhat inconspicuous.

Distribution and abundance: This species is generally rare, reliably recorded from Kuwait and Saudi Arabia (Benzoni 2006). Within the Iranian reefs, it is considered relatively rare and predominantly occupies the shallow reefs surrounding the offshore islands and reef banks towards the Strait of Hormuz in the northern Persian Gulf. It is also recorded in Chabahar Bay in the Gulf of Oman.

Plate 3.162 *Psammocora albopicta* from Larak Island

Plate 3.163 *Psammocora albopicta* from Chabahar Bay

Plate 3.164 Corallites in *Psammocora albopicta* from Larak Island

Plate 3.165 Corallites in *Psammocora albopicta* from Larak Island

Plate 3.166 Corallites in *Psammocora albopicta* from Chabahar Bay

Psammocora contigua (Esper, 1794)

Branched sandpaper coral			
IUCN red list assessment		Status in Iran's reefs	
Status	Population trend	Abundance	Population trend
LC	Decreasing		Decreasing

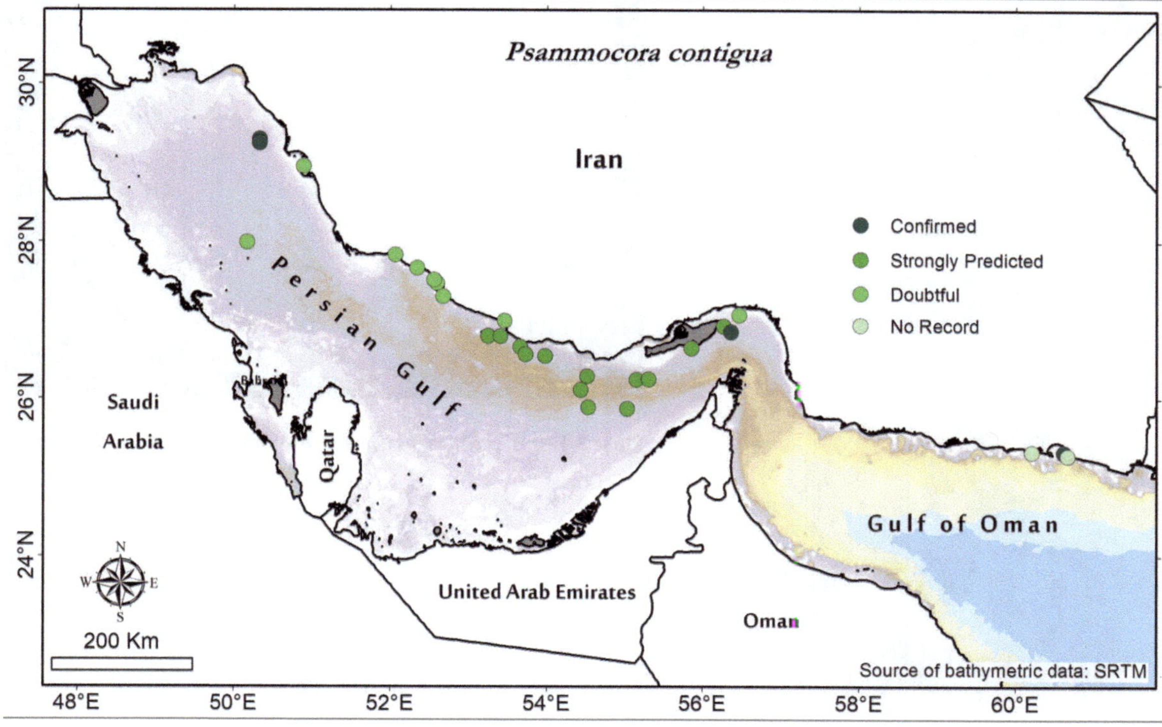

Classification: Phylum—Cnidaria, Class—Hexacorallia, Order—Scleractinia, Family—Psammocoridae

Colony form: Colonies exhibit a mixture of flattened branches, columns, and irregular nodules, occasionally occurring as free-living mobile balls (Plates 3.167 and 3.168). They are usually less than 15 cm in diameter but can grow up to 1 meter. Branches are typically 1 to 2 cm in height, with encrusting forms in low-light conditions. In protected habitats, branches tend to have an angular appearance, while in exposed habitats, they are more rounded. Colonies are often infested with iridescent blue barnacles.

Corallites: The corallite structures are fine and shallow, giving the colonies a smooth surface (Plates 3.169 and 3.170). The corallites are thamnasteroid and irregularly distributed, with up to 10 corallites forming a linear series and subacute collines between them. Corallites range from 1.5 to 4.0 mm in diameter, usually about 2.8 mm, with calices measuring 1.0 to 1.8 mm in diameter. On branches, septa are thin and lamellar with wide intervening spaces, while on flat areas, calices are smaller with thick wedge-shaped septa forming a characteristic petaloid pattern. About 9 to 14 septa reach the fossa, with upper septal margins and septocostae carrying granulated dentations. Fossae are less than 0.5 mm in diameter, and the columella may be a single peg or a cluster of twisted trabeculae. The coenosteum is porous and moderately echinate, with lightly granulated spine tips. *Psammocora contigua* is similar to *P. superficialis* in terms of their fine superficial corallites, but the different growth forms are always distinctive. *P. contigua* is distinguished from *P. superficialis* by its larger, irregularly exsert skeletal elements and angular branches. Slender, tapering tentacles are sometimes extended during the day.

Colour: The color ranges from light to dark brown with a greenish tinge, often lighter at the branch tips.

Distribution and abundance: *Psammocora contigua* generally prefers shallow habitats and soft substrates. It can be found in various environments with a high degree of morphological variation, most commonly at intermediate depths of 3 to 10 meters. This species is recorded from Kuwait (Carpenter et al. 1997). Within the Iranian reefs, it is considered relatively common and predominantly occupies the shallow reefs surrounding the offshore islands and reef banks towards the Strait of Hormuz in the northern Persian Gulf. It is also recorded in Chabahar Bay in the Gulf of Oman.

Plate 3.167 *Psammocora contigua* from Kharg Island

Plate 3.168 *Psammocora contigua* from Larak Island

Plate 3.169 Corallites in *Psammocora contigua* from Kharg Island

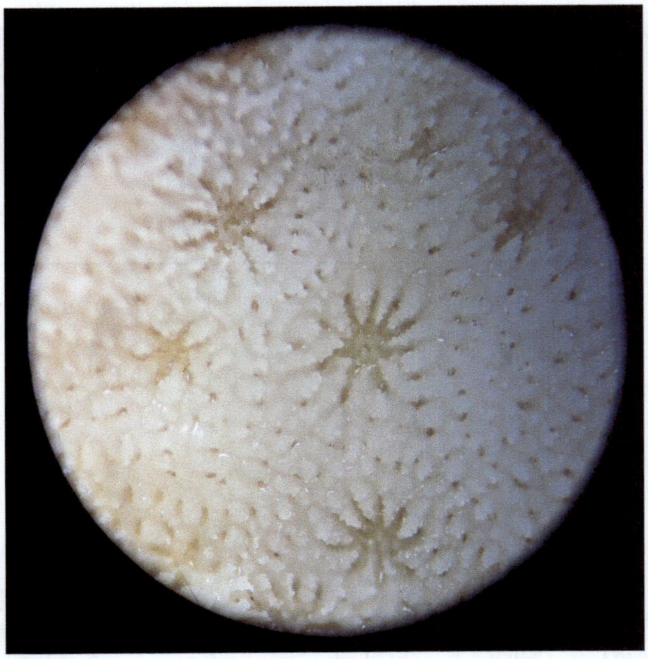

Plate 3.170 Corallites in *Psammocora contigua* from Larak Island

Psammocora columna (Dana, 1846)

Columnar sandpaper coral			
IUCN red list assessment		Status in Iran's reefs	
Status	Population trend	Abundance	Population trend
LC	Decreasing		Decreasing

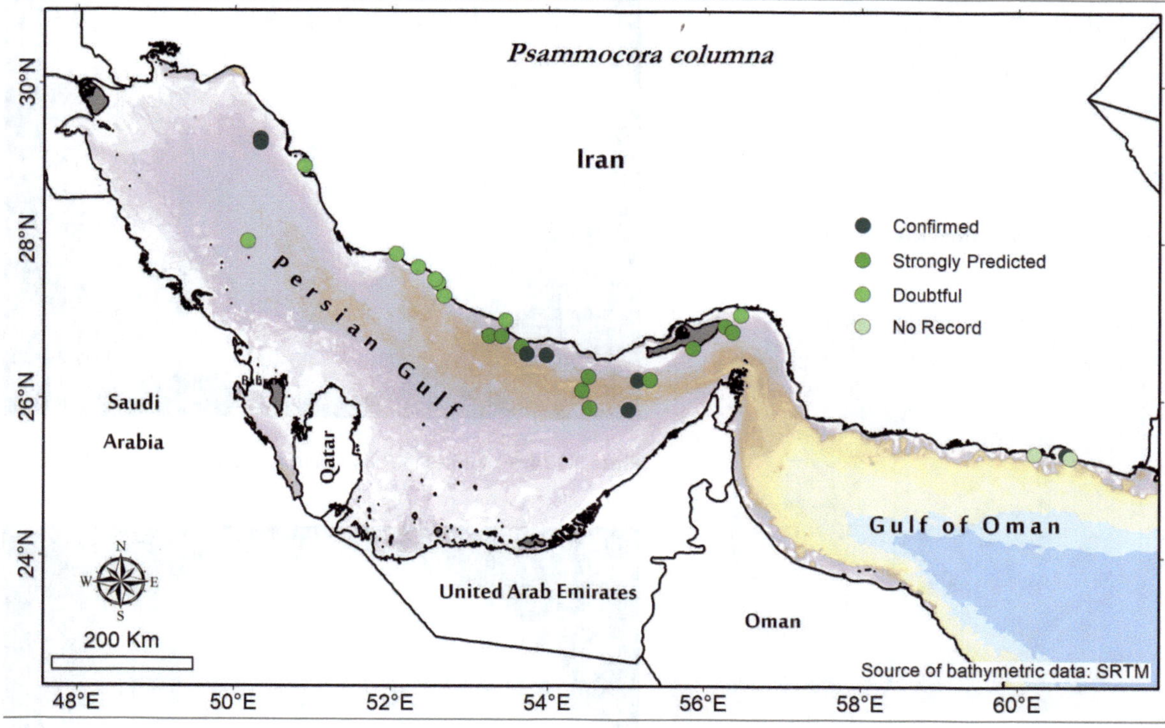

Classification: Phylum—Cnidaria, Class—Hexacorallia, Order—Scleractinia, Family—Psammocoridae

Colony form: Colonies of *P. contigua* exhibit various forms, including encrusting, explanate, or massive structures, typically measuring between 20 to 50 cm in diameter, though occasionally larger specimens are observed (Plates 3.171, 3.172, and 3.173). In shallow marine environments, the colonies display highly developed collines, imparting an irregularly ridged surface appearance. Conversely, in deeper waters, the collines are less pronounced, resulting in a smoother surface texture.

Corallites: The corallites are thamnasteroid, occurring either singly or in linear series of up to 10 corallites within meandroid valleys (Plates 3.174 and 3.175). In certain instances, there is no clear demarcation between adjacent corallites, while in others, a low, thin wall delineates the boundaries. In massive colonies, calices range from 3.0 to 6.0 mm in diameter, whereas in explanate coralla, calices range from 3.0 to 7.0 mm, with an average diameter of approximately 4.5 mm. Up to 40 septa are present at the wall, but only 8 or 9 extend to the columella due to the fusion of adjacent septa. The septa and septocostae exhibit light granulation, with margins divided into sharp dentations. The columellae consist of a tangle of trabeculae or compacted pinnules, measuring approximately 0.5 to 1 mm in diameter. Explanate colonies possess extensive epitheca. Massive colonies of *C. columna* in shallow waters may superficially resemble cerioid to submeandroid forms of *Platygyra daedalea*. The meandroid calices of *P. daedalea* are separated by sharp, high walls, whereas in *C. columna*, the collines are rounded. Additionally, the distinctive brown and white color pattern of *C. columna* may aid in differentiating these species.

Colour: In shallow waters, colonies exhibit brown calices with thick white walls, while in deeper waters, the entire colony assumes a brown coloration.

Distribution and abundance: This species has been reported from Oman and Kuwait (Carpenter et al. 1997). It is commonly found on most Iranian reefs in the Persian Gulf and Gulf of Oman, ranging from the reef flat to the reef base (Plate 3.176).

Plate 3.171 *Psammocora columna* from Kish Island

Plate 3.172 Corallites in *Psammocora columna* from Kharg Island

Plate 3.173 *Psammocora columna* from Sambarun Bank

Plate 3.174 Corallites in *Psammocora columna* from Kish Island

Plate 3.175 Corallites in *Psammocora columna* from Kharg Island

3.3.12 Family Plesiastreidae Dai & Horng, 2009

The Plesiastreidae family is a monotypic family of corals. This family includes only one genus, *Plesiastrea*.

3.3.12.1 Genus *Plesiastrea* Milne Edwards & Haime, 1848

The colonies of *Plesiastrea* are typically massive, either rounded or flattened. The coralla exhibit sub-cerioid to plocoid structures with round corallites formed through budding. Similar to *Montastrea*, *Plesiastrea* has smaller corallites with a well-developed paliform crown composed of true pali. The polyps generally extend only at night. Although *Plesiastrea* is morphologically similar to *Montastraea* (Veron et al., 1977), from an evolutionary perspective, it is more closely related to *Trochocyathus efaatensis* Cairns, 1999 (traditionally classified within the Caryophylliidae Dana, 1846) and *Cyathelia axillaris* (Ellis and Solander, 1786) (currently within the Oculinidae Gray, 1847) than to any other taxon in the Faviidae (Kitahara et al. 2010). *Plesiastrea* is distinguished by the consistent presence of true pali, rather than paliform lobes. Detailed micromorphological analyses (Benzoni et al. 2011) highlight significant differences from other taxa traditionally assigned to the same family. *Plesiastrea* is prevalent in various habitats within the Persian Gulf and Gulf of Oman and is one of the more common species (Coles 1996; Claereboudt 2006; Riegl et al. 2012).

Plate 3.176 Corallites in *Psammocora columna* from Sambarun Bank

Plesiastrea versipora (Lamarck, 1816)

Small knob coral			
IUCN red list assessment		Status in Iran's reefs	
Status	Population trend	Abundance	Population trend
LC	Decreasing		Decreasing

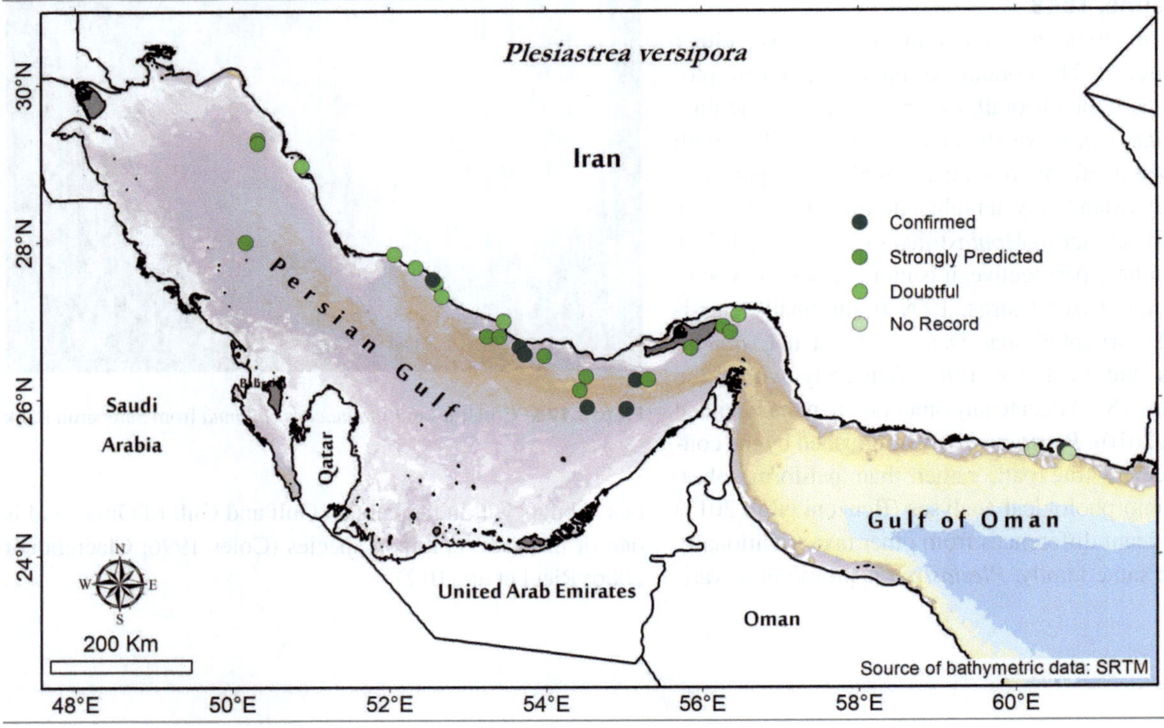

Classification: Phylum—Cnidaria, Class—Hexacorallia, Order—Scleractinia, Family—Plesiastreidae

Colony form: Colonies are encrusting to massive, smooth with low undulations, typically measuring about 15 to 20 cm in diameter, occasionally reaching up to 50 cm (Plate 3.177).

Corallites: Corallites are plocoid, ranging from 3.0 to 5.5 mm in diameter, usually around 4.0 mm (Plate 3.178). Budding is extratentacular. Most corallites are only slightly and evenly exsert above the coenosteum, giving the colony a smooth surface; however, a few widely dispersed corallites may be exsert up to 1.0 mm. Calices possess 20 to 40 septa, typically around 25, regularly arranged in 2 cycles, and evenly exsert. Primary septa arch over the wall and have a sharp indentation before rising to a well-developed palus. The pali form a well-defined crown around the columella. Secondary septa are smaller, denticulate, extending only a short distance from the wall before plunging straight into the calice. Septa and paliform lobes are granulated. Columellae are well-developed, 0.7 to 1.5 mm in diameter, composed of a few pinnacles or vertically twisting trabeculae. In cases where corallites are crowded, they are sharply demarcated by a narrow valley that forms between them. Where corallites are more widely dispersed, the exotheca becomes vesiculate and the costae may extend up to 5 mm out onto the coenosteum. *Plesiastrea* colonies are distinguished from *Cyphastrea* species by having larger corallites with a well-developed, conspicuous paliform crown. The round plocoid corallites of *Plesiastrea* distinguish it from other smooth-surfaced corals such as *Anomastraea*, *Leptastrea*, and *Porites*. Tentacles may be partly extended during the day.

Colour: Colonies are usually bluish-green or green, or dark brown in low-light conditions.

Distribution and abundance: Widespread across the Persian Gulf and Gulf of Oman and present on all Iranian reefs.

Plate 3.177 *Plesiastrea versipora* from Sambarun Bank

Plate 3.178 Corallites in *Plesiastrea versipora*

3.3.13 Family Leptastreidae Rowlett, 2020

The Family Leptastreidae, commonly referred to as crater corals, comprises encrusting or massive corals characterized by medium-sized hexagonal calices with robust septal walls. These corals are widespread in shallow marine environments and are particularly noted for their resilience. Notably, *Leptastrea purpurea* can withstand stressful conditions that are typically fatal to other coral species.

3.3.13.1 Genus *Leptastrea* Milne Edwards & Haime, 1849

Leptastrea colonies are typically massive and exhibit flat or dome-shaped structures. The corallites are subcerioid to plocoid, with small, rounded to polygonal shapes that are often distorted due to crowding, particularly in the Persian Gulf species. The columellae consist of vertical pinnules, while the septa are categorized into two or more size classes, featuring inward-projecting teeth. The well-developed columella comprises rod-like trabeculae, and paliform lobes are not always present, especially in the Gulf species. *Leptastrea* initially classified under the Family Faviidae, it was reassigned to the Family Leptastreidae by Rowlett in 2020 based on phylogenetic studies. This genus, which is closely related to *Cyphastrea*, is characterized by plocoid corallites that are widely separated. Compared to *Cyphastrea*, *Leptastrea* corallites are less exsert, and the costae are either poorly developed or absent. Polyps of *Leptastrea* extend only at night. This genus is represented by five species in the Persian Gulf and Gulf of Oman: *L. purpurea*, *L. transversa*, *L. bottae*, *L.* pruinosa¸ and *L. inaequalis* (Coles 1996; Claereboudt 2006; Riegl et al. 2012). However, only *L. transversa* has been recorded from Iranian reefs in these regions.

Leptastrea transversa **(Klunzinger, 1879)**

Crust coral			
IUCN red list assessment		Status in Iran's reefs	
Status	Population trend	Abundance	Population trend
LC	Decreasing		Decreasing

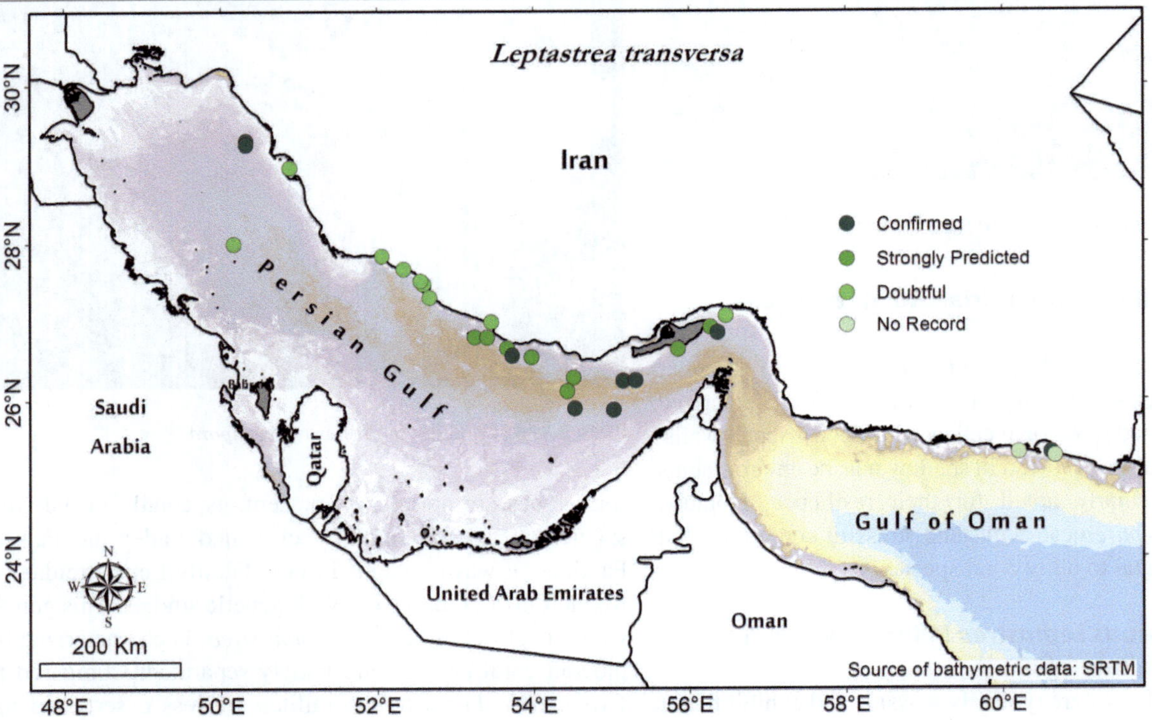

Classification: Phylum—Cnidaria, Class—Hexacorallia, Order—Scleractinia, Family—Leptastreidae

Colony form: Colonies are typically small, ranging from encrusting to submassive forms, with an irregular surface perforated by 1.0 to 2.0 mm holes from iridescent blue barnacles (Plates 3.179 and 3.180). Colony sizes typically range from 20 to 30 cm in diameter, generally less than 50 cm.

Corallites: The corallites are plocoid to sub-cerioid, polygonal, and highly variable in size, measuring between 2 to 7 mm. Budding is extratentacular, and wall thickness is highly variable, up to 2.0 mm thick (Plates 3.181, 3.182, 3.183, and 3.184). The groove separating the calices is narrower and less visible than in *L. purpurea*, but still present. The 30 to 44 septa are organized into three cycles, with the first and second cycles almost equally developed and tending to fuse, while the third cycle is reduced. The costae are poorly developed or absent. The well-developed columella, less than 0.5 mm in diameter, consists of a few trabeculae that are often fused. Septocostae are usually not continuous with those of adjacent corallites, and calices are usually separated by a thin, narrow groove. *L. transversa* is distinguished from *Favites pentagona* by having much smaller corallites, and its colonies never form branches or columns. The corallites of *Anomastraea irregularis* are generally shallower and more regularly organized, giving the colony a smoother, more orderly appearance than *L. transversa*. Brown tentacles are often extended during the day.

Colour: The colonies exhibit colors ranging from green and light blue to brown, with white coenosteum and walls, sometimes with grey-blue centers.

Distribution and abundance: *Leptastrea transversa* has been reported from Oman and Kuwait (Coles 1996; Carpenter et al. 1997). In Iranian reefs, is relatively common on most reefs but is usually not abundant.

3.3.14 Family Coscinaraeidae Benzoni et al. 2012

The Coscinaraeidae family consists of stony corals found primarily in the Indo-Pacific region. These corals are known for their unique skeletal structures. This family includes sev-

Plate 3.179 *Leptastrea transversa* from Kharg Island

Plate 3.180 *Leptastrea transversa* from Larak Island

eral genera, such as *Anomastraea*, *Coscinaraea*, *Craterastrea*, and *Horastrea*. Corallites are found in shallow valleys. The columellae are compact and reach the height of the inner septal margins. These corals are typically uniform in color, often dark grey or brown. Tentacles are often extended during the day.

3.3.14.1 Genus *Anomastraea* Marenzeller, 1901

Anomastraea is a genus of corals within the family Coscinaraeidae, consisting of only one species, the crisp pillow coral (*Anomastraea irregularis*).

Plate 3.181 Corallites in *Leptastrea transversa* from Kharg Island

Plate 3.182 Corallites in *Leptastrea transversa* from Kharg Island

Plate 3.183 Corallites in *Leptastrea transversa* from Larak Island

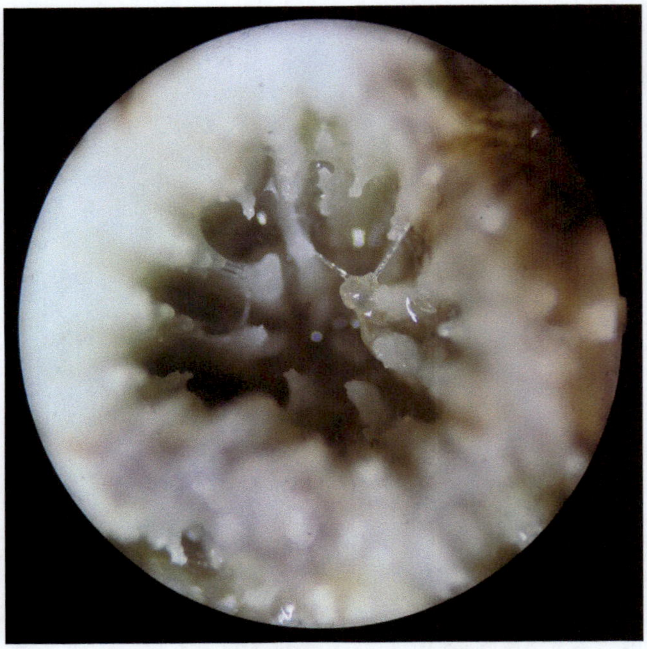

Plate 3.184 Corallite in *Leptastrea transversa* from Larak Island

Anomastraea irregularis (von Marenzeller, 1901)

Crisp pillow coral			
IUCN red list assessment		Status in Iran's reefs	
Status	Population trend	Abundance	Population trend
LC	Decreasing		Decreasing

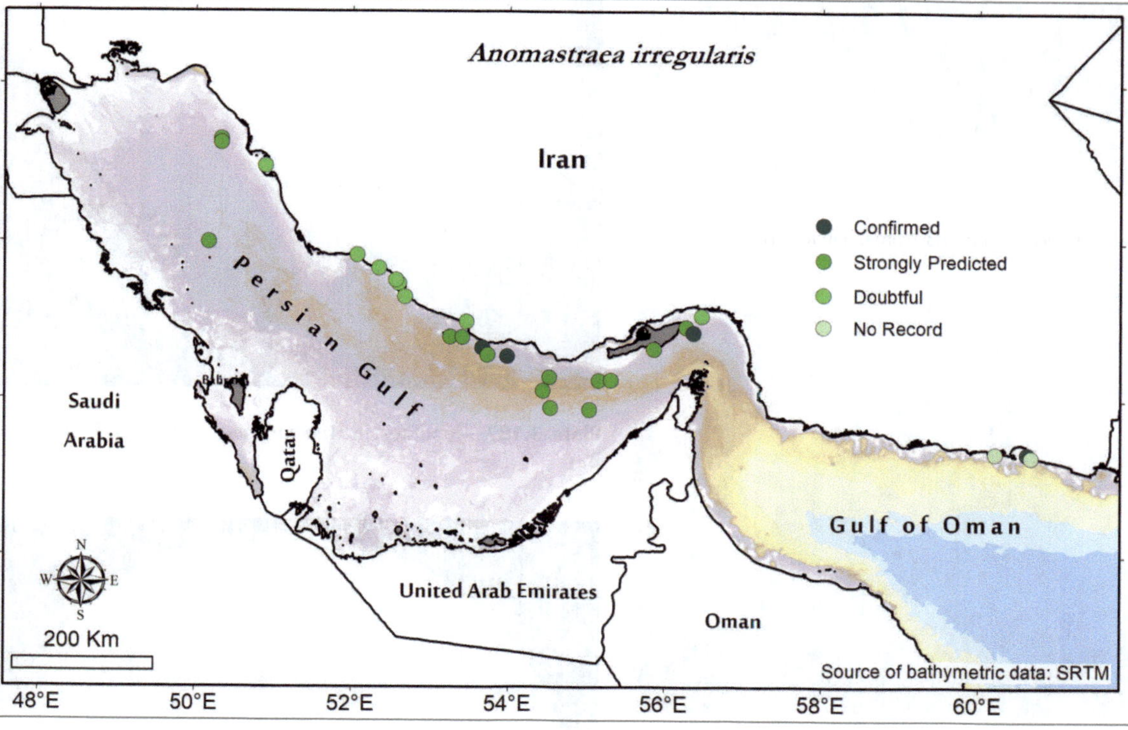

Classification: Phylum—Cnidaria, Class—Hexacorallia, Order—Scleractinia, Family—Coscinaraeidae

Colony form: The corallum can be encrusting, submassive, or massive, with colony widths ranging from 5 to 15 cm and heights between 1 and 12 cm (Plates 3.185 and 3.186).

Corallites: The corallites are cerioid, measuring between 3 and 6 mm in diameter, and are irregularly polygonal, sometimes becoming elongated in certain parts of the colony (Plates 3.187, 3.188, 3.189, and 3.190). The septa number between 15 and 30, depending on the size of the corallite, with no clear distinction into different orders or cycles. Between 1 and 6 septa may reach the columella without fusing, while others fuse at various positions, with more than two septa sometimes fusing. The septa are connected by synapticulae, and their sides are perforated. The columella is present but situated deep in the fossa, composed of processes extending from the septa.

Colour: The coloration of the coral ranges from light brown, yellow, to light green, with the oral discs being light yellow or white.

Distribution and abundance: This coral grows at depths ranging from 5 to 10 meters and is often found in the transitional zone between hard and soft substrate coral communities. This species is commonly found in reef and sandy habitats throughout the Persian Gulf and Gulf of Oman, though it is never abundant.

3.3.14.2 Genus *Coscinaraea* Milne Edwards & Haime, 1848

Colonies of *Coscinaraea* are characterized by their massive, columnar, encrusting, or laminar forms. The corallites are large, situated in short valleys, or irregularly scattered and shallow, with well-developed walls. The columellae consist of groups of pinnules. The septo-costae are fused in distinctive patterns, featuring finely serrated to heavily granulated margins. These corals are common and easily identifiable due to their massive skeletons and large, cerioid polyps. The thecae are well-developed but generally indistinct. Tentacle-bearing septa alternate with exosepta. Although exosepta appear to continue over the wall and become septo-costae, they actually stop on top of the wall where they meet their

Plate 3.185 *Anomastraea irregularis* from Chabahar Bay

Plate 3.186 *Anomastraea irregularis* from Larak Island

counterparts from adjacent corallites. The compound trabeculae and true synapticulothecal wall distinguish *Coscinaraea* from *Psammocora*. Polyps are typically extended at night and sometimes during the day. Eight species of *Coscinaraea* have been reported globally, with two species, *C. monile* and *C. columna*, recorded from the Persian Gulf and Gulf of Oman (Coles 1996; Carpenter et al. 1997; Claereboudt 2006; Riegl et al. 2012). *Coscinaraea monile* is the only species recorded from Iranian reefs in the Persian Gulf and Gulf of Oman.

Plate 3.187 Corallites in *Anomastraea irregularis* from Chabahar Bay

Plate 3.188 Corallites in *Anomastraea irregularis* from Chabahar Bay

Plate 3.189 Corallites in *Anomastraea irregularis* from Larak Island

Plate 3.190 Corallites in *Anomastraea irregularis* from Larak Island

Coscinaraea monile (Forskål, 1775)

Wrinkle coral			
IUCN red list assessment		Status in Iran's reefs	
Status	Population trend	Abundance	Population trend
LC	Decreasing		Decreasing

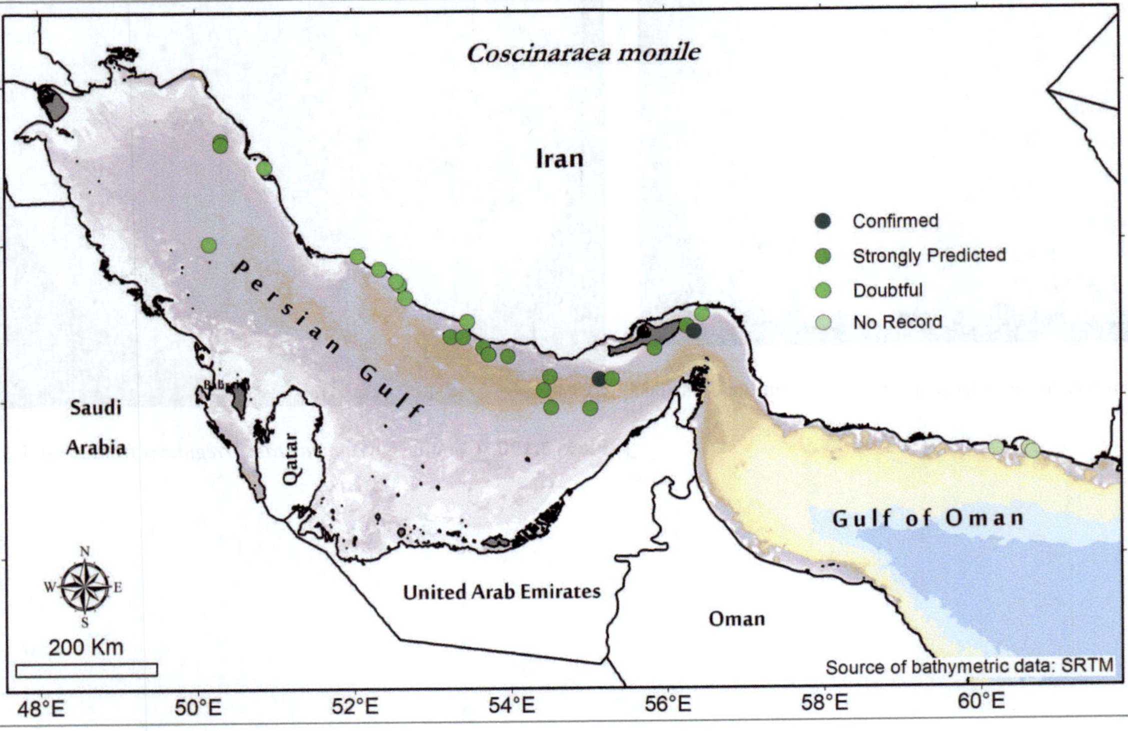

Classification: Phylum—Cnidaria, Class—Hexacorallia, Order—Scleractinia, Family—Coscinaraeidae

Colony form: Colonies of *C. monile* exhibit various forms, including massive, columnar, encrusting, or laminar structures, with free margins in shaded or turbid environments (Plate 3.191).

Corallites: Corallites are large, cerioid, and vary in shape from oval to polygonal or irregular, with diameters ranging from 5 to 15 mm (Plate 3.192). They are typically found in short valleys or scattered irregularly and shallowly. The septa are numerous, around 40, and their number varies between corallites. They feature paddle-shaped ornamentation on the upper edge and are connected by synapticulae, with perforated sides. There is no clear distinction into different orders or cycles; not all septa reach the columella, and many fuse at various positions, with more than two septa sometimes fusing. The columella is present and consists of a tangle of trabeculae.

Colour: The coloration of these corals ranges from beige to light brown. Their typical massive, low growth form and pale coloration make them somewhat inconspicuous and not easily distinguishable.

Distribution and abundance: This coral species is commonly found throughout the Persian Gulf and Gulf of Oman, although it is not abundant in any specific location. It is most noticeable in sandy habitats at the lower end of coral distribution, mainly due to the scarcity of large colonies from genera such as *Porites*, *Platygyra*, and *Stylophora*. Preferring protected or semi-protected habitats, this species can be found at depths ranging from approximately 2–14 meters.

Plate 3.191 *Coscinaraea monile* from Larak Island

Plate 3.192 Corallites in *Coscinaraea monile*

References

Ahmed H, Mohammed H, Ahmed A, Al-Waeli A, Abbas M, Ali M (2023) First record of the stony coral paracyathus stokesii Edward and Haime, 1848 (Caryophylliidae) in the Eastern Breakwater of the New Al-Faw Grand Port, Basrah, Iraq

Aminrad T, Azini MR (2013) New and previous records of scleractinian corals from Chabahar Bay, Sistan & Baluchistan, Iran. Ecopersia 1(4):407–418

Basson P, Burchard JH, Hardy JT, Price A (1977) Biotopes of the prevention, western Arabian Gulf. Aramco Department of Loss and Environmental Affairs, Dhahran

Bauman AG, Pratchett MS, Baird AH, Riegl B, Heron SF, Feary DA (2013) Variation in the size structure of corals is related to environmental extremes in the Persian Gulf. Mar Environ Res 84:43–50. https://doi.org/10.1016/j.marenvres.2012.11.007

Benzoni F (2006) Psammocora albopicta sp nov., a new species of Scleractinian Coral from the Indo-West Pacific (Scleractinia; Siderastreidae). Zootaxa 1358:49–57. https://doi.org/10.11646/zootaxa.1358.1.3

Benzoni F, Stefani F, Stolarski J, Pichon M, Mitta G, Galli P (2007) Debating phylogenetic relationships of the scleractinian Psammocora: Molecular and morphological evidences. Contrib to Zool Bijdr tot Dierkd 76:35–54. https://doi.org/10.1163/18759866-07601004

Benzoni F, Stefani F, Pichon M, Galli P (2010) The name game: Morpho-molecular species boundaries in the genus Psammocora (Cnidaria, Scleractinia). Zool J Linn Soc 160:421–456. https://doi.org/10.1111/j.1096-3642.2010.00622.x

Benzoni F, Arrigoni R, Stefani F, Pichon M (2011) Phylogeny of the coral genus Plesiastrea (Cnidaria, Scleractinia). Contrib to Zool Bijdr tot Dierkd 80:231–249. https://doi.org/10.1163/18759866-08004002

Buchanan JR, Krupp F, Burt JA, Feary DA, Ralph GM, Carpenter KE (2016) Living on the edge: vulnerability of coral-dependent fishes in the Gulf. Mar Pollut Bull 105:480–488. https://doi.org/10.1016/j.marpolbul.2015.11.033

Budd ANNF, Fukami H, Smith ND, Knowlton N (2012) Taxonomic classification of the reef coral family Mussidae (Cnidaria: Anthozoa: Scleractinia). Zool J Linn Soc 166:465–529. https://doi.org/10.1111/j.1096-3642.2012.00855.x

Burchard JE (1979) Coral fauna of the western Persian Gulf. ARAMCO, Dhahran

Burt J, Bartholomew A, Usseglio P (2008) Recovery of corals a decade after a bleaching event in Dubai, United Arab Emirates. Mar Biol 154:27–36. https://doi.org/10.1007/s00227-007-0892-9

Carpenter KE, Harrison PL, Hodgson G, Alsaffar AH, Alhazeem SH (1997) The corals and coral reef fishes of Kuwait. Biological sciences faculty books, 7. https://digitalcommons.odu.edu/biology_books/7

Chevalier JL (1971) Les scleractiniaires de la Melanesie Francaise (Nouvelle Caledonie, Iles Chesterfi eld, Iles Loyaute, Nouvelles Hebrides). 1ere Partie. In: Exped Francaise recifs coralliens Nouvelle Caledonie, vol 5. Fond Singer Polignac. Paris, pp 5–307

Chiappone M, Sealey K, Lott C (1996) Hermatypic scleractinian corals of the Southeastern Bahamas: a comparison to western Atlantic reef systems. Caribb J Sci 32:1–13

Claereboudt MR (2006) Reef Corals and Coral Reefs of the Gulf of Oman. The Historical Association of Oman

Coles SL (1988) Limitations of reef coral development in the Persian Gulf: temperature or algal competition? In: Proceedings of the sixth international coral reef symposium, vol 3, pp 211–216

Coles S (1996) Corals of Oman, p 125

Coles SL (2003) Coral species diversity and environmental factors in the Arabian Gulf and the Gulf of Oman: a comparison to the Indo-Pacific region. Atoll Res Bull:1–19. https://doi.org/10.5479/si.00775630.507.1

Coles SL, Fadlallah YH (1991) Reef coral survival and mortality at low temperatures in the Arabian Gulf: new species-specific lower temperature limits. Coral Reefs 9:231–237. https://doi.org/10.1007/BF00290427

Crabbe MJC, Smith DJ (2005) Sediment impacts on growth rates of Acropora and Porites corals from fringing reefs of Sulawesi,

Indonesia. Coral Reefs 24:437–441. https://doi.org/10.1007/s00338-005-0004-6
DeVantier L, Turak E (2017) Species richness and relative abundance of reef-building corals in the Indo-West Pacific. Diversity 9
Downing N (1995) Coral reef communities in an extreme environment: the Northwest Arabian Gulf. In: Proceedings of the fifth international coral reef symposium, vol 6, pp 343–348
Fadlallah YH, Allen KW, Estudillo RA (1995) Mortality of shallow reef corals in the western Arabian Gulf following aerial exposure in winter. Coral Reefs 14:99–107. https://doi.org/10.1007/BF00303430
Fatemi SMR, Shokri MR (2001) Iranian coral reefs status with particular reference to Kish Island, Persian Gulf. In: Indian Ocean regional workshop, p 13
Fukami H, Chen CA, Budd AF, Collins A, Wallace C, Chuang Y-Y, Chen C, Dai C-F, Iwao K, Sheppard C, Knowlton N (2008) Mitochondrial and nuclear genes suggest that stony corals are monophyletic but most families of stony corals are not (Order Scleractinia, Class Anthozoa, Phylum Cnidaria). PLoS One 3:e3222
Ghazilou A (2019) Bleaching of corals in Chah Bahar Bay, Sea of Oman. Tehran
Goffredo S, Dubinsky ZVY (2016) The Cnidaria, past, present and future: the world of Medusa and her sisters. Springer International Publishing
Grandcourt E (2012) In: Riegl BM, Purkis SJ (eds) Reef fish and fisheries in the Gulf BT – Coral Reefs of the Gulf: adaptation to climatic extremes. Springer, Dordrecht, pp 127–161
Harger JRE (1984) Rapid survey techniques to determine distribution and structure of coral communities. In: Comparing coral reef survey methods. UNEPUNESCO Workshop, Thailand. Phuke, Thiland, pp 83–91
Hodgson G, Carpenter K (1995) Scleractinian corals of Kuwait. Pacific Sci 49:227–246
Hoeksema BW, Cairns S (2024a) world List of Scleractinia. Fungiidae Dana, 1846. Accessed through: World Register of Marine Species at: https://www.marinespecies.org/aphia.php?p=taxdetails&id=196100
Hoeksema BW, Cairns S (2024b) World list of Scleractinia. Oxypora Saville Kent, 1871. Accessed through: World Register of Marine Species at: https://www.marinespecies.org/aphia.php?p=taxdetails&id=205536
Kavousi J, Seyfabadi J, Rezai H, Fenner D (2011) Coral reefs and communities of Qeshm Island, the Persian Gulf. Zool Stud 50:276–283
Kitahara MV, Cairns SD, Stolarski J, Blair D, Miller DJ (2010) A comprehensive phylogenetic analysis of the scleractinia (Cnidaria, Anthozoa) based on mitochondrial CO1 sequence data. PLoS One 5:e11490
Kitano YF, Benzoni F, Arrigoni R, Shirayama Y, Wallace CC, Fukami H (2014) A phylogeny of the Family Poritidae (Cnidaria, Scleractinia) based on molecular and morphological analyses. PLoS One 9:e98406
Maghsoudlou A (2008) Hard Corals of the Iranian coastal waters of the Persian Gulf. Iranian National Institute for Oceanography and Atmospheric Science (INIOAS)
Maghsoudlou A (2020) Check list of Iran marine Cnidarians (Animalia, Cnidaria). Iran J Anim Biosyst 16:1–12. https://doi.org/10.22067/ijab.v16i1.85363
Miller KJ (1992) Morphological variation in the scleractinian coral Platygyra daedalea (Ellis and Solander, 1786), genetically or environmentally determined? In: Proceedings of the seventh international coral reef symposium, vol 1, pp 550–556
Miller K, Babcock R (1997) Conflicting morphological and reproductive species boundaries in the coral genus platygyra. Biol Bull 192:98–110. https://doi.org/10.2307/1542579
Miller K, Benzie J (1997) no clear genetic distinction between morphological species within the coral genus Platygyra. Bull Mar Sci 61:907–917
Moradi M, Malekzadeh-Viayeh R, Eshaghi-Rad J (2014) Biodiversity of scleractinian corals in the reefs of Qeshm and Larak Islands of the Persian Gulf, in association with environmental variables. J Mar Biol Assoc United Kingdom 94:907–916. https://doi.org/10.1017/S0025315414000411
Orrell T (2019) NMNH extant specimen records. Version 1.21. National Museum of Natural History, Smithsonian Institution. Occurrence dataset.
Pratchett M, Baird A, Bauman A, Burt J (2016) Abundance and composition of juvenile corals reveals divergent trajectories for coral assemblages across the United Arab Emirates. Mar Pollut Bull. https://doi.org/10.1016/j.marpolbul.2016.11.036
Purkis S, Riegl B (2005) Spatial and temporal dynamics of Arabian Gulf coral assemblages quantified from remote-sensing and in situ monitoring data. Mar Ecol Ser 287:99–113. https://doi.org/10.3354/meps287099
Purkis SJ, Riegl BM (2012) Geomorphology and reef building in the SE Gulf. In: Riegl BM, Purkis SJ (eds) BT coral reefs of the Gulf: adaptation to climatic extremes. Springer, Dordrecht, pp 33–50
Rahmani M, Rahimian H (2013) Preliminary study on Acropora (Scleractinia: Astrocoeniina: Acroporidae) of the Persian Gulf, with emphasis on the north and northeastern areas. Turkish J Zool 37:308–320. https://doi.org/10.3906/zoo-1112-21
Rezai H, Wilson S, Claereboudt M, Riegl B (2004) Coral reef status in the ROPME sea area: Arabian/Persian Gulf, Gulf of Oman and Arabian Sea. Status of Coral Reefs of the World, pp 155–170.
Riegl B (1999) Corals in a non-reef setting in the southern Arabian Gulf (Dubai, UAE): fauna and community structure in response to recurring mass mortality. Coral Reefs 18:63–73. https://doi.org/10.1007/s003380050156
Riegl B (2001) Inhibition of reef framework by frequent disturbance: examples from the Arabian Gulf, South Africa, and the Cayman Islands. Palaeogeogr Palaeoclimatol Palaeoecol 175:79–101. https://doi.org/10.1016/S0031-0182(01)00387-X
Riegl B (2003) Climate change and coral reefs: different effects in two high-latitude areas (Arabian Gulf, South Africa). Coral Reefs 22:433–446. https://doi.org/10.1007/s00338-003-0335-0
Riegl BM, Purkis SJ (2012a) Coral reefs of the Gulf: adaptation to climatic extremes in the World's Hottest Sea. In: Riegl BM, Purkis SJ (eds) BT – coral reefs of the Gulf: adaptation to climatic extremes. Springer, Dordrecht, pp 1–4
Riegl BM, Purkis SJ (2012b) Environmental constraints for reef building in the Gulf. In: Riegl BM, Purkis SJ (eds) BT – coral reefs of the Gulf: adaptation to climatic extremes. Springer, Dordrecht, pp 5–32
Riegl BM, Purkis SJ, Al-Cibahy AS, Abdel-Moati MA, Hoegh-Guldberg O (2011) Present limits to heat-adaptability in corals and population-level responses to climate extremes. PLoS One. https://doi.org/10.1371/journal.pone.0024802
Riegl BM, Benzoni F, Samimi-Namin K, Sheppard C (2012) The hermatypic scleractinian (hard) coral fauna of the Gulf. In: Coral reefs of the Gulf: adaptation to climatic extremes. Springer, Dordrecht, pp 187–224
Riegl B, Johnston M, Purkis S, Howells E, Burt J, Steiner SCC, Sheppard CRC, Bauman A (2018) Population collapse dynamics in Acropora downingi, an Arabian/Persian Gulf ecosystem-engineering coral, linked to rising temperature. Glob Chang Biol 24:2447–2462. https://doi.org/10.1111/gcb.14114
Sadeghi P, Loghmani M (2009) New observation of three species of hard coral from Chabahar Bay (Oman Sea), Iran. J Biol Sci 9:837–840. https://doi.org/10.3923/jbs.2009.837.840
Sale PF, Feary DA, Burt JA, Bauman AG, Cavalcante GH, Drouillard KG, Kjerfve B, Marquis E, Trick CG, Usseglio P, Van Lavieren H (2011) The growing need for sustainable ecological management of marine communities of the Persian Gulf. Ambio 40:4–17. https://doi.org/10.1007/s13280-010-0092-6

Salimi PA, Chen PGMCA, Fatemi SMR, Pichon M (2018) The Scleractinia (Cnidaria: Anthozoa) of Abu-Musa and Sirri Islands, Persian Gulf

Samimi-Namin K, Ofwegen L (2009) Some shallow water octocorals (Coelenterata: Anthozoa) of the Persian Gulf. Zootaxa:1–52. https://doi.org/10.5281/zenodo.186743

Sharif Ranjabr M, Askari Hesni M, Ghavasi M, Dab K, Chegonian A (2015) Integrated coral reef management program of Tonb, Abu Musa, and Siri Islands by hard corals species identification, health status control and distribution mapping. Tehran

Sheppard CRC (1988) Similar trend, different causes: responses of corals to stressed environments in Arabian seas. In: Proceedings of the 6th international coral reef congress, Townsville, vol 3, pp 297–302

Sheppard C (2016) Coral reefs in the Gulf are mostly dead now, but can we do anything about it? Mar Pollut Bull 105:593–598. https://doi.org/10.1016/j.marpolbul.2015.09.031

Sheppard C, Salm R (1988) Reef and coral communities of Oman, with a description of a new coral species (Order Scleractinia, genus Acanthastrea). J Nat Hist 22:263–279. https://doi.org/10.1080/00222938800770201

Sheppard CR, Sheppard ALS (1991) Corals and coral communities of Arabia. In: Buttiker W, Krupp F (eds) Fauna of Saudi Arabia, vol 12. Natural History Museum, Basle, pp 3–170

Sheppard CRC, Sheppard ALS (1993) Corals and coral communities of Arabia. In: Fauna of Saudi Arabia, pp 12, 170

Sheppard C, Price A, Roberts C (1992) Marine ecology of the Arabian Region: patterns and processes in extreme tropical environments, 1st edn. Academic Press

Sheppard C, Al-Husiani M, Al-Jamali F, Al-Yamani F, Baldwin R, Bishop J, Benzoni F, Dutrieux E, Dulvy NK, Durvasula SRV, Jones DA, Loughland R, Medio D, Nithyanandan M, Pilling GM, Polikarpov I, Price ARG, Purkis S, Riegl B, Saburova M, Namin KS, Taylor O, Wilson S, Zainal K (2010a) The Gulf: a young sea in decline. Mar Pollut Bull 60:13–38. https://doi.org/10.1016/j.marpolbul.2009.10.017

Sheppard C, Al-Husiani M, Al-Jamali F, Al-Yamani F, Baldwin R, Bishop J, Benzoni F, Dutrieux E, Dulvy N, Durvasula S, Jones D, Loughland R, Medio D, Manickam N, Pilling G, Polikarpov I, Price A, Purkis S, Riegl B, Zainal K (2010b) The Persian/Arabian Gulf: a young sea in decline. Mar Pollut Bull 60:13–38. https://doi.org/10.1016/j.marpolbul.2009.10.017

Shinn EA (1976) Coral reef recovery in Florida and the Persian Gulf. Environ Geol 1:241–254. https://doi.org/10.1007/BF02407510

Shojae F, Kamrani E, Shokri MR, Ranjbar MS, Moradi M, Hesni MA (2010) New records of three hard coral species from north-east of Larak Island (Persian Gulf, Iran). Mar Biodivers Rec 3:1–6. https://doi.org/10.1017/s1755267210000515

Shokri MR (2001) Status of coral reefs around farur island, iran (persian gulf) final report 2001, pp 1–35

Shokri MR, Mohammadi M (2021) Effects of recreational SCUBA diving on coral reefs with an emphasis on tourism suitability index and carrying capacity of reefs in Kish Island, the northern Persian Gulf. Reg Stud Mar Sci 45:101813. https://doi.org/10.1016/j.rsma.2021.101813

Shokri MR, Oladi M (2021) New record of azooxanthellate dendrophyllia cribrosa in the Persian gulf (Anthozoa, scleractinia, dendrophylliidae). Spixiana 44:11–16

Shokri MR, Ardakani O, Sharifi A, Abdoullahi P, Nazarian M (2000) Status of coral reefs around Kish Island in the Iranian Persian Gulf. In: Proceedings of the international symposium on the extent and impact of coral bleaching in the Arabian Region, pp 5–9

Uchupi E, Swift SA, Ross DA (1996) Gas venting and late Quaternary sedimentation in the Persian (Arabian) Gulf. Mar Geol 129:237–269. https://doi.org/10.1016/0025-3227(96)83347-0

Vajed Samiei J, Dab K, Ghezellou P, Shirvani A (2013) Some scleractinian corals (Scleractinia: Anthozoa) of Larak Island, Persian Gulf. Zootaxa 3636:101–143. https://doi.org/10.11646/zootaxa.3636.1.5

Vaughan TW, Wells JW (1943) Revision of the suborders families, and Genera of the Scleractinia. Geological Society of America

Venkataraman K, Satyanarayana C (2012) Coral Identification manual. Zoological Survey of India, Kolkata

Veron JEN (1986) Corals of Australia and the Indo-Pacific. Angus and Robertson, Sydney

Veron JEN (2000) Corals of the world. Australian Institute of Marine Science, Townsville

Veron JEN, Pichon M (1976) Scleractinia of Eastern Australia. Australian Govt. Pub. Service, Canberra

Veron JEN, Pichon M, Wijsman-Best M (1977) Scleractinia of Eastern Australia, II: families Faviidae, Trachyphylliidae. AIMS Monogr Ser 3:227

Veron JEN, John EN, Wallace CC (1984) Scleractinia of eastern Australia. Part V. Family Acroporidae. Scleractinia East Aust Part V 6:1–485

Vogt H (1996) Investigations on coral reefs in the Jubail wildlife sanctuary using under water video recordings and digital image analysis. In: Krupp F, Abuzinada AH, Nader IA (eds) A marine wildlife sanctuary for the Persian Gulf. NCWCD Riyadh and Senckenbergische Naturforschende Gesellschaft, Frankfurt, pp 302–327

Wallace CC (1978) The coral genus Acropora (Scleractinia: As_trocoeniina; Acroporidae) in the central and southern Great Barrier Reef. Mem Queensl Museum 18:273–319

Wallace C (1999) Staghorn corals of the world. CSIRO Publishing

Weil E, Hammerman NM, Becicka RL, Cruz-Motta JJ (2020) Growth dynamics in Acropora cervicornis and A. prolifera in southwest Puerto Rico. PeerJ 2020:1–23. https://doi.org/10.7717/peerj.8435

Wells JW (1956) Scleractinia. In: Moore R (ed) Treatise on invertebrate paleontology. Part F. Coelenterata. University Kansas Press, Lawrence, pp 328–444

Wilson S, Fatemi SMR, Shokri MR, Claereboudt M (2002) Status of coral reefs of the Persian/Arabian Gulf and Arabian Sea Region. Status of coral reefs of the world, pp 53–62

WoRMS Editorial Board (2004) World register of marine species. In: VLIZ. https://www.marinespecies.org

Yap HT, Gomez ED (1984) Growth of Acropora pulchra. Mar Biol 81:209–215. https://doi.org/10.1007/BF00393119

4 Conservation and Management of Iran's Coral Reefs

4.1 Coral Reefs in Iranian Waters: Threats and Conservation Imperatives

4.1.1 Coastal Development

Unregulated coastal development is one of the most immediate and severe threats to Iran's coral reefs. As coastal populations increase and tourism expands, infrastructure projects—such as ports, resorts, and residential complexes—are proliferating. Large-scale developments like the Pars Special Economic Energy Zone (PSEEZ) in the Persian Gulf promote industrial activity in oil, gas, and petrochemical sectors, which necessitate extensive dredging, land reclamation, and the construction of artificial waterways. These modifications result in significant alterations to coastal and marine ecosystems.

In areas such as Kish and Qeshm islands, extensive land reclamation and construction activities have led to habitat destruction and elevated sedimentation rates. Sediment plumes from dredging can smother coral colonies, reduce light penetration critical for photosynthesis, and inhibit coral larval recruitment. Over time, these effects lead to diminished coral cover and reduced reef resilience.

Environmental impact assessments (EIAs), where conducted, are frequently inadequate or poorly enforced, resulting in unmitigated ecological harm. Additionally, the removal of mangrove forests—natural buffers that trap sediments and filter pollutants—further compounds reef stress in adjacent areas.

4.1.2 Marine Pollution

Marine and land-based pollution are major threats to the health of coral reef ecosystems in Iran. Key sources include untreated sewage, industrial discharges, oil spills, and agricultural runoff (Plate 4.1). These inputs introduce a wide range of harmful substances into reef environments, including excess nutrients, heavy metals, and petroleum hydrocarbons. Elevated nutrient levels, for instance, promote the overgrowth of macroalgae, which compete with corals for space and sunlight. Meanwhile, toxic pollutants impair coral physiology, reproduction, and immune responses, further weakening reef resilience.

A significant contributor to pollution in the region is the growing number of desalination plants, especially along the northern coast of the Persian Gulf. These facilities discharge hypersaline brine into coastal waters, increasing local salinity and disrupting the osmotic balance essential to coral survival. Studies conducted around Kish Island have documented elevated concentrations of contaminants in reef-associated sediments and marine organisms, indicating long-term exposure to pollutants from desalination and other sources.

Marine debris, including plastic waste, abandoned fishing gear, and general litter, also poses a threat to coral reefs, albeit to a lesser extent than other stressors in Iran (Plate 4.2). Debris can physically smother corals, block sunlight necessary for photosynthesis, and damage delicate reef structures through entanglement. Moreover, leaching chemicals from waste introduce additional toxins into the ecosystem, exacerbating environmental stress. Although this threat

M. R. Shokri, K. Kabiri, *Coral Reefs of Iran*, Coral Reefs of the World 22, https://doi.org/10.1007/978-3-032-00598-4_4

Plate 4.1 Increased sedimentation from coastal runoff (Kish Island) diminishes water clarity, disrupts coral photosynthesis, and heightens susceptibility to disease

is considered relatively minor compared to others, it remains a concern that warrants continued management and public awareness.

In light of these challenges, sustained conservation efforts are essential to mitigate the impacts of marine pollution. Addressing both point and non-point sources of pollution through improved waste management, stricter regulations, and increased public engagement will be key to safeguarding Iran's coral reefs for future generations.

4.1.3 Overfishing and Destructive Fishing Practices

Overfishing, including the use of destructive and illegal fishing gear, poses another significant threat to coral reefs in Iranian waters. The removal of herbivorous fish disrupts algal control mechanisms, enabling algae to outcompete coral polyps. The depletion of predatory species further destabilizes food webs.

In Iran, illegal gear such as wire mesh traps (Gargoor) and fine-mesh nets damages reef structures and reduces biodiversity (Plates 4.3 and 4.4). The collection of ornamental marine species, including corals and seahorses for the aquarium trade, also exerts additional pressure on reef ecosystems (Plate 4.5). Reef-associated fish stocks, especially around heavily populated islands, are exhibiting signs of overexploitation.

Improper diving practices, such as touching and trampling on coral reefs, cause severe physical damage, increase the reefs' vulnerability to disease, and disrupt the overall stability of the reef ecosystem, as observed in Kish, Qeshm, and Larak Islands—key hubs for recreational SCUBA diving (Plate 4.6). Additionally, spearfishing poses a major threat to reef fish populations, leading to imbalances in the marine food web and contributing to ecosystem degradation, as seen in both inland and offshore Iranian reefs (Plate 4.7).

Plate 4.2 Marine debris, including plastic waste and abandoned fishing gear, threatens coral reef ecosystems by causing physical damage, smothering corals, and introducing harmful pollutants into the environment

4.1.4 Climate Change

Climate change is a global driver of coral reef degradation, and Iranian reefs are increasingly affected (Bargahi et al. 2020; Shokri 2020; Lincoln et al. 2021; Maltby et al. 2022). Rising sea surface temperatures (SSTs) have triggered mass coral bleaching events, including major episodes in 1998, 2010, and 2017. During bleaching, corals expel their symbiotic zooxanthellae, leading to energy depletion and, potentially, mortality.

Although Persian Gulf corals exhibit some degree of thermal tolerance, SSTs regularly exceed 34 °C during summer, pushing corals to their physiological limits (Plate 4.8). The increasing frequency and intensity of marine heatwaves could surpass their adaptive capacities. Furthermore, ocean

Plate 4.5 The collection of ornamental marine species, including corals and seahorses for the aquarium trade, contributes to reef ecosystem degradation by increasing exploitation pressures on fragile marine habitats

Plate 4.3 An abandoned wire mesh trap (Gargoor) resting on *Acropora* corals in Kish Island. (September 2007, Photo by H. Valavi)

Plate 4.6 Improper diving practices can harm coral reefs, with touching and trampling causing physical damage, increased vulnerability to disease, and disruption of reef stability. (Kish Island, March 2019)

Plate 4.4 Corals smothered by an abandoned gill net at Khargoo Island. (January 2007, Photo by H. Valavi)

Plate 4.7 Spear fishing poses a significant threat to reef fish populations. (Nayband Bay, April 2025)

acidification, resulting from increased atmospheric CO_2, inhibits coral calcification, compromising reef accretion and structural integrity.

4.1.5 Oil Extraction and Maritime Activities

The Persian Gulf is a central hub for oil production and maritime transport. Oil spills, accidental discharges, and chronic hydrocarbon pollution from ships and offshore platforms introduce toxins into marine environments. Coral reefs near oil terminals and shipping lanes—such as those around Kharg, Lavan, and Siri islands—are particularly vulnerable.

Oil pollution impairs gas exchange and light penetration, crucial for coral metabolism. Sublethal concentrations of hydrocarbons interfere with coral reproductive processes and larval development. In some cases, the use of chemical dispersants during oil spill cleanups exacerbates ecological damage by introducing additional pollutants.

Maritime activities, including anchoring (Plate 4.9) and ballast water discharge, further contribute to coral degradation. Anchoring in shallow reef areas causes physical damage to coral colonies and disrupts the reef framework. Repeated disturbances prevent natural recovery. Additionally, ballast water introduces non-native species and pathogens, increasing ecological risk.

4.1.6 Sedimentation and Turbidity

Elevated sedimentation from coastal runoff, construction, and dredging activities reduces water clarity, impairs coral photosynthesis, and increases disease susceptibility (Plate 4.10). Fine sediments can settle on coral tissue, causing smothering and abrasion. Suspended particles hinder larval settlement and recruitment, limiting coral population recovery.

Climate-related phenomena, such as increased storm intensity, can exacerbate sediment input, particularly during monsoonal events (Plate 4.11). These episodic stressors compound chronic impacts from human activities.

4.1.7 Invasive Species and Coral Diseases

Although still understudied in Iran, invasive species and coral diseases are increasingly recognized as emerging threats to reef health. Elevated sea surface temperatures and nutrient enrichment promote the onset and spread of disease outbreaks. Pathogenic bacteria, particularly *Vibrio* spp., have been associated with coral tissue necrosis and bleaching events. Documented coral diseases in Iranian reefs include Arabian Yellow Band (AYB) disease, pink spot and pink line diseases, as well as skeletal growth anomalies (Fatemi and Shokri 2001; Tavakoli-Kolour et al. 2015; Salimi et al. 2017) (Plates 4.12 and 4.13).

Non-native species introduced via ballast water or through aquarium releases may disrupt local reef ecosystems by competing with native species or introducing novel pathogens. The lack of monitoring and response mechanisms leaves Iranian reefs vulnerable to such invasions.

4.1.8 Cumulative and Synergistic Threats to Iran's Coral Reef Ecosystems

Coral reefs in Iran provide vital ecological functions as well as economic and cultural benefits to coastal communities. However, these ecosystems are under increasing pressure from a combination of human-induced stressors. The cumulative impact of these threats is not merely additive; their synergistic nature amplifies ecological degradation, severely weakening reef resilience and diminishing the potential for natural recovery.

In the Persian Gulf, the most severe threats to coral reefs, in descending order of impact, include coastal development, fishing activities, desalination and industrial discharges, dredging and shipping operations, sewage and wastewater pollution, and tourism-related disturbances. In contrast, the Gulf of Oman faces a different set of priorities: coastal development poses the greatest threat, followed by fishing, tourism and recreation, desalination and industrial discharges, and sewage and wastewater pollution (Plate 4.14).

Despite the urgency of these threats, current management approaches remain fragmented and largely reactive. Institutional coordination among government agencies, research institutions, and local communities is limited, and the lack of comprehensive long-term monitoring and baseline data continues to hinder the development of effective conservation policies.

To address these challenges, adaptive management strategies are essential—approaches that integrate scientific research with meaningful community engagement. Such strategies can help mitigate ongoing losses and build reef resilience over time.

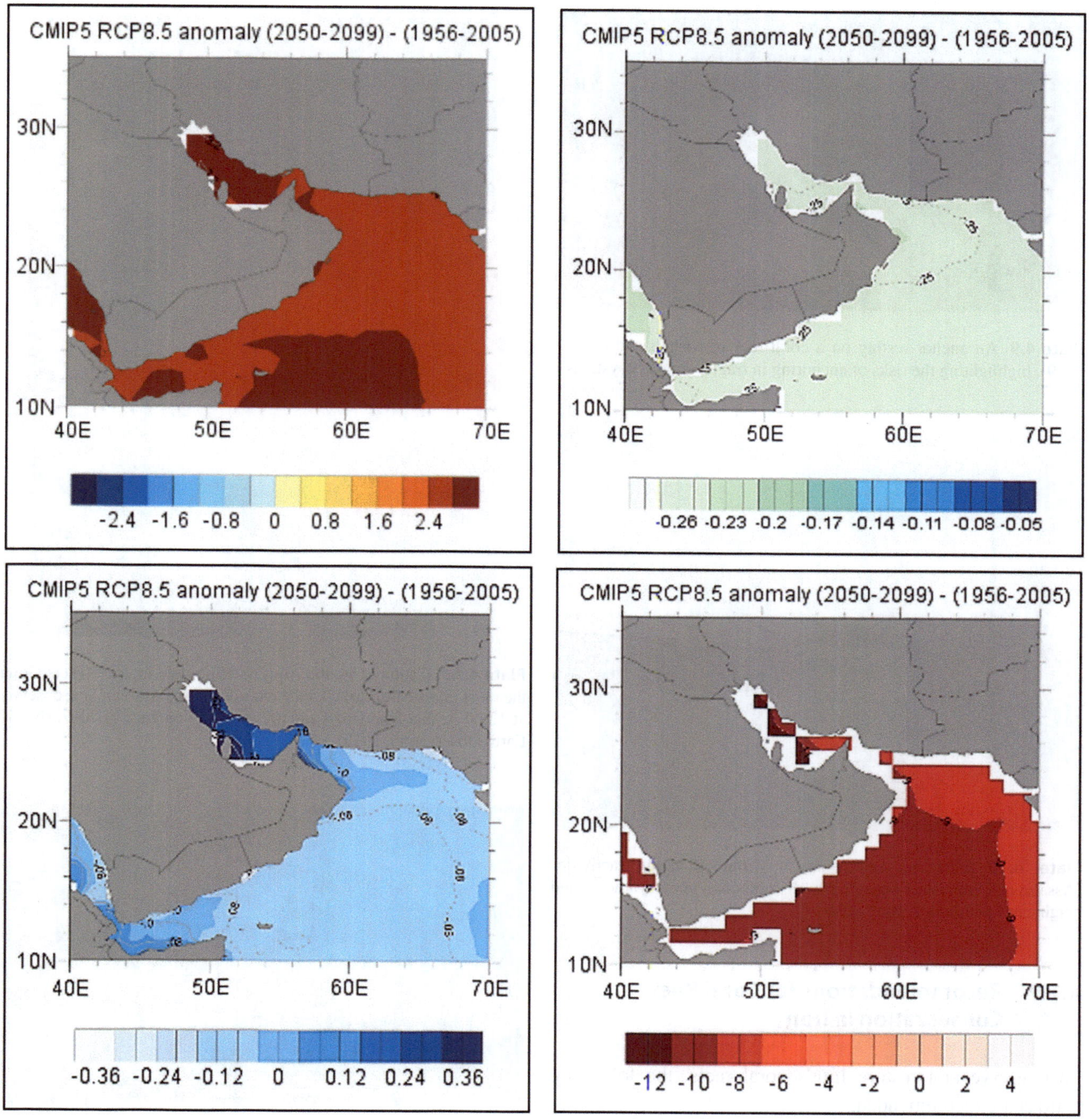

Plate 4.8 Projected changes in key oceanographic parameters in the Persian Gulf and Gulf of Oman under the RCP 8.5 scenario (2050–2099), compared to the historical baseline (1956–2005), based on CMIP5 models. The maps illustrate variations in annual average sea surface temperature (top left), pH (top right), salinity (bottom left), and dissolved oxygen (bottom right). White shading indicates areas with no available data. (ESRL: PSD: Climate Change Web Portal, Maps MM, 2018)

The discussion concludes with an open call for participants to share insights from their monitoring activities. Emphasis is placed on highlighting successful interventions and lessons learned from past experiences. This collaborative knowledge exchange is key to enhancing coral reef conservation efforts in Iran and developing more effective, inclusive strategies moving forward.

Plate 4.9 An anchor resting on a coral reef in Kish Island (March 2019), highlighting the risks of anchoring in fragile coral ecosystems

Plate 4.10 Increased sedimentation from dredging activities (Assaluyeh) diminishes water clarity, disrupts coral photosynthesis, and heightens susceptibility to disease

4.1.9 Recommendations for Coral Reef Conservation in Iran

To conserve and restore Iran's coral reefs, the following actions are recommended:

- **Establish and effectively manage Marine Protected Areas (MPAs)** with adequate funding and enforcement mechanisms.
- **Enforce stringent regulations** on coastal development, dredging, and pollution control.
- **Promote sustainable fisheries management**, including the prohibition of destructive gear and provision of alternative livelihoods for fishers.
- **Enhance coral resilience** through active restoration efforts, such as coral gardening and assisted evolution research.

Plate 4.11 Captured by the Aqua satellite, this image depicts Gonu, the most intense tropical cyclone recorded in the Arabian Sea, observed at 12:00 Arabia Standard Time (09:00 UTC) on June 4, 2007. (NASA Earth Observatory, 2019)

Plate 4.12 Yellow Band Disease on massive coral *Porites harrisoni*. (Kish Island, March 2017)

- **Develop and implement early warning systems** for bleaching, disease outbreaks, and pollution events.
- **Foster regional collaboration** with Gulf nations for coordinated management and data sharing.
- **Increase public awareness and environmental education** about the importance of coral reef conservation.

In sum, Iran's coral reefs are at a critical juncture. Without immediate and sustained conservation action, these ecosystems face the risk of functional collapse. An integrated, multidisciplinary approach is essential to address the complex environmental, socio-economic, and governance challenges. Investing in science-based management, stakeholder engagement, and regional cooperation can ensure the long-term sustainability of Iran's coral reefs for future generations.

4.1.10 Conservation Strategies and Initiatives in Iran: Safeguarding Coral Reef Biodiversity

Coral reefs in Iran face a host of anthropogenic and climate-related threats including coastal development, pollution, sedimentation, destructive fishing, and warming waters. These pressures necessitate a robust set of conservation strategies to ensure the persistence and resilience of coral reef ecosystems. Iran has initiated several local and national conservation efforts—ranging from the designation of Marine Protected Areas (MPAs) and coral restoration projects to sustainable resource management practices—aimed at curbing reef degradation and preserving marine biodiversity.

Plate 4.13 Pink discoloration and irregular markings on coral indicate stress responses to physical and pathogenic factors, including fish bites, parasites, burrowing invertebrates, and microbial infections. (Kish Island, March 2017)

4.1.10.1 Marine Protected Areas (MPAs)

One of the cornerstone strategies for coral reef conservation in Iran is the establishment of Marine Protected Areas (MPAs). These are legally designated zones where human activities are regulated to conserve the ecological integrity of the marine environment. Iran has declared multiple MPAs that include coral reef habitats:

- **Nayband Marine National Park (NMNP)** in the Bushehr Province stands out as Iran's first marine national park. Encompassing over 40,000 hectares, it includes coral communities, mangroves, seagrass beds, and important nesting areas for sea turtles and seabirds. Its designation reflects both ecological and socio-economic considerations.
- **Dayyer-Nakhiloo MPA**, which includes Nakhiloo Island and adjacent waters, is another example of an area providing sanctuary to coral assemblages as well as critical habitats for hawksbill turtles and migratory birds.
- **Qeshm Island Geopark** (a UNESCO-listed site) and **Hengam Island**, both in the Strait of Hormuz, are managed for biodiversity conservation and ecotourism, though management enforcement is variable.

Although the creation of Marine Protected Areas (MPAs) marks a notable advancement, evidence indicates that numerous sites continue to struggle with enforcement, resource distribution, and stakeholder involvement. The oversight, monitoring, and regulation of human activities—particularly illegal fishing, coastal development, and tourism—remain insufficient.

Moreover, there is an **urgent need for an integrated network of MPAs** that ensures ecological connectivity between coral reef habitats. Presently, the fragmented nature of MPAs hinders the resilience of coral metapopulations, particularly as reefs experience increasing stress from climate change.

Plate 4.14 Key challenges impacting coral reefs in Iran's Persian Gulf and Gulf of Oman, ranked by severity

4.1.10.2 Coral Restoration Projects

Given the widespread coral mortality reported in the region—primarily due to elevated sea surface temperatures, sedimentation, and disease—coral restoration has become an important conservation tool in Iran.

Pilot-scale coral restoration projects have been initiated in several locations, notably:

- **Kish Island**: Restoration efforts here have included coral gardening and transplantation techniques. Fragments of opportunity and nursery-grown corals are transplanted onto degraded reef patches.
- **Larak and Qeshm islands**: These sites have hosted small-scale community-driven restoration initiatives, with involvement from local NGOs, divers, and students. These projects often focus on fast-growing species like *Acropora* and *Porites*, with the dual aim of enhancing biodiversity and engaging the public.
- **Kharg Island**: In collaboration with the oil industry, efforts have been made to transplant corals from pipline construction-impacted areas to safe reef locations, minimizing biodiversity loss due to industrial activities.

However, coral restoration remains **experimental and poorly documented**, with limited long-term monitoring (Darkhal et al. 2025). The restoration should **complement, not substitute**, broader conservation policies such as pollution control and habitat protection.

4.1.11 Sustainable Management Practices

Iran's conservation strategy also includes a suite of sustainable management practices targeting the root causes of reef degradation. These are focused on both terrestrial and marine sources of stress and emphasize ecosystem-based management:

4.1.11.1 Fisheries Regulation

Overfishing and destructive fishing practices, such as the use of trawls and nets near reef habitats, have severely impacted coral ecosystems. Sustainable fisheries management efforts include:

- Seasonal bans on shrimp and demersal fisheries.
- Regulation of gear types and mesh sizes.
- Restrictions on fish catch for certain species, including herbivores and reef predators.

Although these regulations exist on paper, **enforcement remains inconsistent**, especially in remote islands and smaller fishing communities. There is also a lack of species-specific stock assessments to guide adaptive management.

4.1.11.2 Pollution Control

Effluent discharge, oil pollution, and plastic waste are major threats to Iran's reefs, particularly near industrial hubs like Bandar Abbas and Assaluyeh. Initiatives to manage land-based pollution include:

- Installation of wastewater treatment plants in urban areas like Qeshm and Kish islands.
- Regular monitoring of water quality parameters, especially in tourist-heavy zones.
- Regulations for oil terminals and offshore platforms, including mandatory contingency plans for oil spills.

However, **illegal discharge and sedimentation from coastal construction** (e.g., land reclamation and port development) continue to erode coral resilience. This highlights the urgent need for stricter environmental impact assessments (EIAs) and a more coordinated enforcement framework.

4.1.11.3 Public Awareness and Education

Iran has invested in **environmental education programs** targeting schools, universities, and coastal communities. These efforts promote stewardship, ecotourism, and citizen science initiatives such as reef monitoring. For example:

- University students are often engaged in reef surveys and mapping exercises.
- NGOs conduct beach cleanups and educational outreach on plastic pollution and marine conservation.
- Interpretive centers on Kish and Qeshm islands serve as hubs for public learning about marine biodiversity.

Expanding on these initiatives, adopting more structured, community-led management strategies—such as the collaborative governance of MPAs or locally coordinated enforcement patrols—could enhance stakeholder involvement and boost overall effectiveness.

4.1.12 National and Institutional Framework

Iran's coral reef conservation is supported by a national environmental governance framework involving several key institutions:

- **Department of Environment (DoE)**: Leads national conservation policy, protected area management, and biodiversity assessments.
- **Iranian Fisheries Organization (IFO)**: Oversees fisheries regulation and marine resource use.
- **Iranian National Institute for Oceanography and Atmospheric Science (INIOAS)**: Conducts marine research and monitoring, including coral reef assessments.

Despite these efforts, there is **need for improved coordination between these bodies**, especially to avoid overlapping mandates and fragmented implementation. A national action plan for coral reef conservation, aligned with the Convention on Biological Diversity (CBD) and Sustainable Development Goals (SDGs), is recommended to unify efforts.

Additionally, regional cooperation under the **Regional Organization for the Protection of the Marine Environment (ROPME)** offers a platform for shared data, joint monitoring, and regional restoration strategies.

4.1.13 Challenges and Recommendations

Iran faces several challenges in effectively conserving its coral reefs. Climate change has led to elevated sea surface temperatures, triggering repeated coral bleaching events. To address this, local conservation efforts must be integrated into broader global climate mitigation frameworks. Additionally, inadequate funding and capacity remain significant obstacles, as many marine protected areas (MPAs) lack trained personnel, necessary infrastructure, and long-term financial support. The absence of continuous, standardized reef health monitoring further hampers conservation efforts, limiting the ability to track long-term trends and adjust management strategies accordingly. Another critical issue is limited stakeholder involvement; top-down approaches frequently exclude key groups such as fishers, tourism operators, and local communities, weakening compliance and long-term stewardship.

To overcome these challenges, several key recommendations can be implemented. Scaling up MPAs into a connected and resilient network will help protect ecological corridors and ensure the long-term survival of coral ecosystems. Coral restoration should be mainstreamed into coastal development planning, with incentives for industry compliance to encourage sustainable practices. Strengthening legal and institutional frameworks, particularly enforcement capacity, is essential to ensure conservation regulations are effectively implemented. A more integrated approach that combines scientific research with traditional knowledge will allow for adaptive management models that respond to environmental changes. Additionally, enhancing regional collaboration, especially for transboundary reef ecosystems, will create opportunities for shared data, joint monitoring initiatives, and coordinated conservation strategies.

In conclusion, Iran's coral reefs are ecologically rich but increasingly vulnerable ecosystems. While national and local conservation strategies show promise, they must evolve into more cohesive, enforceable, and participatory frameworks to ensure long-term effectiveness. The combined efforts of MPAs, coral restoration, and sustainable management practices provide a strong foundation for coral reef conservation, but success ultimately depends on sustained political will, active community engagement, scientific innovation, and regional cooperation. As marine pressures continue to escalate, particularly under climate change, the window for safeguarding Iran's coral biodiversity is narrowing. Taking decisive action now is crucial to securing the future of these vital ecosystems.

4.1.14 Policy and Global Collaboration: Protecting Iran's Coral Reefs Through National Strategy and International Partnership

Iran's coral reef ecosystems, situated in the Persian Gulf and Gulf of Oman, are ecologically vital and increasingly vulnerable. These ecosystems support rich marine biodiversity and serve as critical habitats, fisheries grounds, and coastal protection structures. However, they are under severe stress from anthropogenic pressures such as coastal development, pollution, oil extraction, and the broader impacts of climate change.

Against this backdrop, Iran has developed marine biodiversity policies and engaged in regional and global collaborations to fulfill its environmental responsibilities. Notably, international conventions such as the **Convention on Biological Diversity (CBD)** and the **United Nations Sustainable Development Goals (SDGs)** have served as guiding frameworks. This summary synthesizes Iran's policy landscape and its role in global and regional efforts aimed at coral reef preservation, drawing from the 2020 technical report and contextualizing it within international obligations.

4.1.14.1 National Marine Biodiversity Policy Framework

Iran's policies on marine biodiversity are primarily administered through the **Department of Environment (DoE)**, in coordination with the **Iranian Fisheries Organization (IFO)** and research institutions like the **Iranian National Institute for Oceanography and Atmospheric Science (INIOAS)**. Key features of Iran's biodiversity policy include:

4.1.14.2 Protected Area Designation

Iran has established a system of **Marine Protected Areas (MPAs)**, many of which encompass coral reef habitats. Examples include Nayband Marine National Park, Larak Island, and parts of Qeshm Geopark. These MPAs are grounded in national legislation aiming to preserve biodiversity and ecosystem services.

4.1.14.3 Environmental Impact Assessment (EIA) Regulations

Major coastal and offshore development projects—especially those related to oil and gas—require EIAs. These assessments are mandated to consider impacts on coral ecosystems. However, the report notes that enforcement and monitoring are uneven, with frequent lapses in post-construction compliance.

4.1.14.4 National Biodiversity Strategies and Action Plans (NBSAPs)

In compliance with the CBD, Iran has developed **National Biodiversity Strategies and Action Plans**, which identify coral reefs as priority ecosystems. These plans emphasize integrated coastal zone management (ICZM), pollution control, and stakeholder engagement.

4.1.14.5 Marine Pollution and Habitat Protection

Policies regulating industrial effluents, maritime oil transport, and marine litter have been introduced. The **Marine Pollution Law** governs discharge from ships and offshore platforms, though implementation remains variable.

While these policies demonstrate Iran's commitment to biodiversity conservation, they face persistent challenges, including:

- Fragmented institutional responsibilities.
- Limited financial and technical resources.
- Insufficient integration of scientific research into policy decisions.

4.1.14.6 Iran and the Convention on Biological Diversity (CBD)

Iran has been a **Party to the CBD since 1996**, committing the country to biodiversity conservation, sustainable use, and equitable benefit-sharing. The CBD provides the foundational structure for Iran's biodiversity governance.

4.1.14.7 Key CBD Commitments Reflected in National Policy

- **Aichi Biodiversity Targets (2011–2020):** Iran has made efforts to address several of the 20 Aichi targets, especially those related to protecting ecosystems, reducing pollution, and promoting traditional knowledge.
- **Post-2020 Global Biodiversity Framework:** Although the national response is still evolving, the report highlights the importance of aligning future policy with new targets focusing on ecosystem restoration and resilient MPA networks.

4.1.14.8 Coral Reefs in the NBSAP

Iran's NBSAP recognizes coral reefs as a priority for conservation. Strategies include:

- Expanding MPAs to cover ecologically representative reef habitats.
- Promoting habitat restoration through coral transplantation projects.
- Integrating coral monitoring into national biodiversity databases.

Despite these efforts, there is **need for measurable indicators**, improved inter-agency coordination, and stronger engagement with local communities and NGOs.

4.1.14.9 Sustainable Development Goals (SDGs) and National Alignment

Iran has pledged to support the **UN's 2030 Agenda for Sustainable Development**, which includes 17 interconnected goals. Several SDGs are directly relevant to coral reef protection:

4.1.14.10 Relevant SDGs

- **SDG 14 (Life Below Water):** Calls for sustainable management and protection of marine and coastal ecosystems.
- **SDG 13 (Climate Action):** Addresses climate change adaptation and resilience-building.
- **SDG 6 (Clean Water and Sanitation):** Encourages reduction of marine pollution through better wastewater treatment.

4.1.14.11 National Implementation

Iran's national development plans integrate SDG principles. For instance:

- Programs promoting wastewater infrastructure in coastal provinces like Hormozgan and Bushehr.
- Pilot coral restoration and reef monitoring initiatives.
- Use of MPAs as SDG indicators for conservation progress.

However, **there is limited disaggregated data** on SDG indicators at the local reef level. Enhanced reporting mechanisms are needed to assess progress toward marine sustainability goals more transparently.

4.1.14.12 Regional and International Collaboration

In addition to global frameworks, Iran participates in **regional initiatives** and **bilateral scientific collaborations** that bolster coral reef conservation.

4.1.14.13 ROPME (Regional Organization for the Protection of the Marine Environment)

Iran is a founding member of ROPME, a regional body coordinating marine protection among Persian Gulf states. ROPME provides:

- **Shared monitoring protocols** for coral reef health and pollution levels.
- **Early warning systems** for oil spills and algal blooms.
- A **platform for data exchange** and policy alignment.

Despite ROPME's commendable role in fostering regional solidarity, challenges persist due to political tensions and limitations in data sharing.

4.1.14.14 UNEP Regional Seas Programme

Through its participation in the **Kuwait Action Plan**, Iran contributes to UNEP's goal of regional marine protection. This includes:

- Periodic assessments of marine biodiversity.
- Technical workshops on coral monitoring.
- Capacity-building for EIA processes and marine spatial planning.

4.1.14.15 International Scientific Research

Iranian institutions have collaborated with researchers from Japan, Australia, and European countries on:

- Coral bleaching surveys using remote sensing.
- Climate-resilience studies of coral genotypes.
- Genetic assessments of reef fish and invertebrates.

These collaborations help compensate for gaps in local capacity, particularly in the use of molecular and geospatial tools.

4.1.14.16 Gaps and Recommendations

Despite progress, Iran's coral reef policy and global engagement face several **strategic gaps**:

4.1.14.17 Policy Enforcement and Monitoring

While policies exist, their enforcement is weak due to limited human and financial resources. The report calls for:

- Increased investment in MPA infrastructure and staff training.
- Use of satellite and drone technologies for compliance monitoring.

4.1.14.18 Integration of Science into Policy

Scientific findings are not always translated into management decisions. A formal mechanism is needed to:

- Integrate coral health assessments into EIA and MPA planning.
- Institutionalize long-term reef monitoring programs.

4.1.14.19 Community Involvement and Traditional Knowledge

Top-down policies frequently disregard the perspectives of local stakeholders, including fishers and tourism operators. To address this, we should advocate for co-management approaches in marine protected areas (MPAs), ensuring that local communities play an active role in decision-making. Additionally, integrating traditional ecological knowledge into reef conservation efforts can enhance sustainability and strengthen the connection between conservation initiatives and local practices.

4.1.14.20 Climate Change Adaptation

Most current policies are reactive. Proactive climate adaptation strategies, such as identifying climate refugia and protecting resilient coral species, are lacking.

4.1.14.21 Stronger International Engagement

Iran's scientific and policy engagement in global platforms (e.g., IPBES, ICRI) is limited. Active participation could bring:

- Access to international funding (e.g., Green Climate Fund).
- Technical support for coral restoration and marine spatial planning.

Iran's policies for coral reef conservation are evolving within a framework shaped by national priorities and international obligations. Through its commitment to the CBD and the SDGs, Iran has articulated the importance of marine biodiversity and laid the foundation for reef protection through MPAs, pollution control, and scientific research.

Yet, implementation remains uneven, and there is a need for more integrated, well-funded, and participatory approaches. Strengthening ties with global and regional partners—especially through ROPME and UNEP—will be crucial for meeting conservation goals. As coral reefs face escalating threats, Iran must scale up its efforts and embrace innovative, science-driven, and community-centered strategies to safeguard its underwater heritage.

References

Bargahi HR, Shokri MR, Kaymaram F, Fatemi MR (2020) Changes in reef fish assemblages following multiple bleaching events in the world's warmest sea (Kish Island, the Persian Gulf). Coral Reefs 39:603–624. https://doi.org/10.1007/s00338-020-01945-3

Darkhal H, Kabiri K, Shokri MR (2025) Optimizing coral reef restoration: A GIS-based site selection for coral nurseries in Kish Island, the Persian Gulf. J Nat Conserv 86:126898. https://doi.org/10.1016/j.jnc.2025.126898

Fatemi SMR, Shokri MR (2001) Iranian coral reefs status with particular reference to Kish Island, Persian Gulf. In: Indian Ocean regional workshop, p 13

Lincoln S, Buckley P, Howes EL, Maltby KM, Pinnegar JK, Ali TS, Alosairi Y, Al-ragum A, Baglee A, Balmes CO, Ben HR, Burt JA, Claereboudt M, Glavan J, Mamiit RJ, Abdulla H, Naser A, Sedighi O, Shokri MR, Shuhaibar B (2021) A regional review of marine and coastal impacts of climate change on the ROPME sea area

Maltby KM, Howes EL, Lincoln S, Pinnegar JK, Buckley P, Ali TS, Al Balushi B, Al Ragum A, Al Shukail HSA, Balmes CO, Ben-Hamadou R, Claereboudt MRG, Mamiit RJE, Naser HA, Shokri MR, Le Quesne WJF (2022) Marine climate change risks to biodiversity and society in the ROPME Sea Area. Clim Risk Manag 35:100411. https://doi.org/10.1016/j.crm.2022.100411

Salimi MA, Mostafavi PG, Fatemi SMR, Aeby GS (2017) Health status of corals surrounding Kish Island, Persian Gulf. Dis Aquat Organ 124:77–84. https://doi.org/10.3354/dao03105

Shokri MR (2020) Status and trends of coral reefs in the ROPME sea area: past, present and future. Kuwait

Tavakoli-Kolour P, Kavousi J, Rezai H (2015) Outbreak of growth anomalies in coral communities of Qeshm Island, Persian Gulf. Int Aquat Res 7:151–156. https://doi.org/10.1007/s40071-015-0100-3

Methods for Monitoring Coral Reefs

5

5.1 Remote Sensing and Satellite Technologies

Remote sensing has become a cornerstone of coral reef monitoring in Iran, particularly in the Persian Gulf, where physical access is often limited (Plate 5.1). Satellite-based technologies provide large-scale, repeatable assessments of coral reef distribution, benthic composition, and changes in reef extent over time. High spatial resolution platforms such as QuickBird, WorldView series (1–4), and GeoEye offer valuable multispectral imagery capable of distinguishing live coral cover from algae, sand, or rubble. These technologies are widely employed in the northern Persian Gulf, especially around Kish, Qeshm, and Hengam islands, where reefs are highly vulnerable to sedimentation, coastal development, and thermal stress.

Time-series analysis of satellite data has proven valuable in assessing reef degradation and monitoring the impacts of mass bleaching events, especially those associated with extended sea surface temperature anomalies. Additionally, geographic information systems (GIS) integrated with remote sensing are utilized to map reef extent and model environmental threats, including land-based pollution and shoreline morphology changes. These combined tools empower Iranian marine scientists and policymakers to pinpoint key areas for conservation and restoration efforts.

5.2 Drone-Based Reef Monitoring

In recent years, drone technology (Unmanned Aerial Vehicles—UAVs) has emerged as a game-changing tool for coral reef monitoring in Iran, particularly around islands like Kish, Larak, and Qeshm (Plate 5.2). Their affordability, agility, and ability to capture ultra-high-resolution aerial imagery make them ideally suited for mapping and studying shallow-water reef systems. In recent studies conducted near Kish Island, UAVs were utilized to generate detailed orthomosaics and integrate aerial photographs with depth data from single-beam sonar, resulting in 3D models of reef structures. These outputs have been instrumental in analyzing spatial patterns of coral cover, detecting bleaching events, and identifying localized threats such as sedimentation or algal overgrowth.

Drones have also facilitated non-invasive surveys in turbid or logistically challenging conditions, minimizing physical disturbance to delicate reef habitats. Additionally, these studies have highlighted the potential of engaging local stakeholders—such as university students and community groups—in UAV-based monitoring programs, promoting scientific participation and conservation awareness. By combining drone imagery with machine learning techniques, these efforts are advancing the automated classification of benthic habitats, significantly improving the accuracy, efficiency, and reproducibility of ecological assessments."

5.3 Ecological Surveys and Data Analysis

Despite technological advancements, in situ ecological surveys remain indispensable for capturing the complexity of coral reef ecosystems. Along the Iranian coastline, monitoring efforts adhere to standard protocols such as Line Intercept Transects (LIT) for measuring coral cover and benthic composition, Belt Transects for assessing species richness and abundance, and Quadrat sampling for evaluating community structure at finer scales (Plate 5.3).

Since 1999, the implementation of ReefCheck methodology has facilitated extensive fieldwork across Nayband Bay, Kish, Qeshm, and Larak islands, generating long-term datasets crucial for assessing reef health trends. These surveys, conducted using SCUBA and snorkel-based methods, enable divers to document reef-associated fauna, signs of disease, and the presence of invasive species.

The collected data undergo rigorous analysis using biodiversity indices such as Shannon-Wiener and Pielou's evenness, alongside multivariate statistical techniques like cluster analysis and Non-Metric Multidimensional Scaling (NMDS).

M. R. Shokri, K. Kabiri, *Coral Reefs of Iran*, Coral Reefs of the World 22, https://doi.org/10.1007/978-3-032-00598-4_5

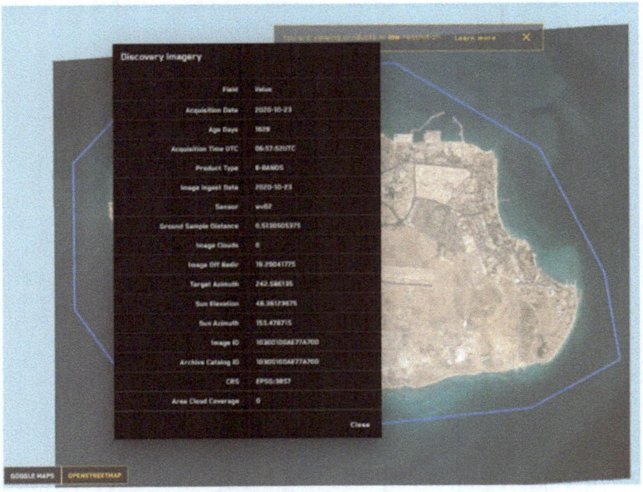

Plate 5.1 Sample satellite image of Kish Island, evaluated for coverage and quality (e.g., cloud cover) prior to ordering

Increasingly, R-based ecological modeling tools provide deeper insights into coral community dynamics, resilience to environmental stressors, and the effectiveness of conservation strategies.

Although coral reef monitoring in Iran has advanced, particularly over the past decade, its continuity remains challenged by reliance on external funding. Some ROPME Member States maintain their own monitoring programs or participate in international coral reef networks such as GCRMN and ReefCheck. However, unstable funding sources, personnel changes, and institutional shifts result in discontinuities within national datasets, compromising the consistency and long-term value of collected data. Furthermore, inconsistencies in monitoring methodologies across different States hinder regional and national reporting efforts, reducing the applicability of data for policy-making, environmental awareness, and scientific research. Addressing these challenges is vital to ensure reliable reef monitoring that can effectively guide conservation and management decisions.

Plate 5.2 Drone-based coral reef mapping near Qeshm Island, northern Persian Gulf. (Top-left) Flight planning interface for the DJI Phantom 4 Pro mission conducted on 5 January 2025, showing parameters such as altitude (50 m), GSD (1.4 cm), and coverage area. (Top-right) Satellite image pinpointing the study site along the southern coast of Qeshm Island. (Bottom-left) Field operation depicting the drone operator during takeoff. (Bottom-right) High-resolution orthomosaic image of the reef and nearshore zone, created from aerial photographs taken during the flight

Plate 5.3 Monitoring the reefs of Kish Island using the ReefCheck methodology (March 2019)

6 Future Directions in Coral Reef Research

6.1 Emerging Research Areas

Coral reef research in Iran is entering a transformative era, driven by climate change and increasing human pressure. One major research frontier involves g**enetic and molecular approaches** to understand coral resilience (Moghaddam et al. 2021). These reseraches initial steps toward **genotyping coral species** to identify thermally tolerant genotypes, particularly in environments with extreme salinity and temperature such as the Persian Gulf.

Symbiotic interactions, especially between corals and their algal partners (*Symbiodiniaceae*), are another focal area. Researchers are investigating how changes in symbiont composition affect coral stress responses. Moreover, ocean acidification, a growing concern, is being studied for its impact on **calcification processes**, skeletal density, and coral reproduction.

Additionally, **microbiome studies** are gaining traction, exploring the role of bacterial communities in coral health and disease resistance. These studies integrate metagenomics and functional profiling to uncover beneficial microbes that could be harnessed for coral probiotics or assisted evolution strategies.

6.2 Technological Innovations in Coral Monitoring

A surge in smart reef monitoring is anticipated, driven by advancements in emerging technologies. Some key innovations include:

- **Artificial Intelligence (AI) and Deep Learning**: Being applied to automate classification of coral cover and detect anomalies (e.g., bleaching or sedimentation) from drone and underwater imagery.
- **Genetic Barcoding and Environmental DNA (eDNA)**: Offering precise species identification, even from small tissue samples or waterborne genetic material, thus facilitating biodiversity assessments and early detection of invasive species.
- **Autonomous Underwater Vehicles (AUVs) and Remotely Operated Vehicles (ROVs)**: Enabling high-resolution surveys at greater depths where traditional divers cannot reach. These are particularly promising for Iran's lesser-explored mesophotic reefs.
- **Real-time Environmental Monitoring Systems**: Using sensors to continuously record parameters like temperature, pH, dissolved oxygen, and turbidity, which are essential for predictive models of reef health.

These innovations hold great promise for enhancing the accuracy, frequency, and depth of reef monitoring—critical in the face of rapid environmental changes.

6.3 Collaborative Global Efforts

Iran's coral reef ecosystems, though localized, are part of a broader regional and global network affected by shared challenges, including **climate change, marine pollution, and overfishing**. The report strongly advocates for **strengthened regional cooperation**, particularly among Gulf countries, to ensure effective conservation.

We suggest deeper involvement in international frameworks such as:

- **International Coral Reef Initiative (ICRI)**
- **UNEP Regional Seas Programme**
- **Global Coral Reef Monitoring Network (GCRMN)**

Participation in global reef restoration and monitoring projects can offer Iranian scientists access to standardized meth-

M. R. Shokri, K. Kabiri, *Coral Reefs of Iran*, Coral Reefs of the World 22, https://doi.org/10.1007/978-3-032-00598-4_6

odologies, shared databases, and funding mechanisms, enhancing research and conservation efforts. Bilateral collaboration with neighboring countries that share coral reef ecosystems, including Kuwait, Saudi Arabia, Bahrain, Qatar, the UAE, and the Sultanate of Oman, as well as leading coral research nations such as the United States and Australia, could strengthen local capacity building and drive innovation in reef conservation.

Science diplomacy (Fawzi et al. 2022; Fieseler et al. 2023) in the Gulf is essential, with coral reef conservation offering a unique opportunity to foster environmental collaboration beyond political differences. Advancing this effort through joint publications, shared research expeditions, and data-sharing agreements can pave the way for a more cohesive and effective regional conservation strategy.

References

Fawzi NA-M, Fieseler CM, Helmuth B, Leitão A, Al-Ainsi M, Al Mukaimi M, Al-Saidi M, Al Senafi F, Bejarano I, Ben-Hamadou R, D'Addario J, Mohamed AMD, Giraldes BW, Glowka L, Johnson MD, Lyons BP, Mateos-Molina D, Marshall CD, Mohammed S, Range P, Shokri MR, Wong JMK, Pyenson ND (2022) Diplomacy for the world's hottest sea. Science 376:1389–1390. https://doi.org/10.1126/science.add1555

Fieseler CM, Al-Mudaffar Fawzi N, Helmuth B, Leitão A, Al Ainsi M, Al Mukaimi M, Al-Saidi M, Al Senafi F, Bejarano I, Ben-Hamadou R, D'Addario J, Mujthaba Dheen Mohamed A, Giraldes BW, Glowka L, Johnson MD, Lyons BP, Mateos-Molina D, Marshall CD, Mohammed S, Range P, Reza Shokri M, Wong JMK, Pyenson ND (2023) Expanding ocean protection and peace: a window for science diplomacy in the Gulf. R Soc Open Sci 10:230392. https://doi.org/10.1098/rsos.230392

Moghaddam S, Shokri MR, Tohidfar M (2021) The enhanced expression of heat stress-related genes in scleractinian coral 'Porites harrisoni' during warm episodes as an intrinsic mechanism for adaptation in 'the Persian Gulf'. Coral Reefs 40:1013–1028. https://doi.org/10.1007/s00338-021-02100-2

Glossary

Aerial Photogrammetry The process of using aerial photographs to measure and map features on the Earth's surface, commonly used in coral reef mapping through drone-acquired images.

Agisoft™ Metashape Photogrammetric processing software used to generate orthophoto mosaics and 3D models from overlapping drone images.

Ahermatypic Corals Non-reef-building corals that typically lack zooxanthellae and can inhabit deeper or colder waters, relying on heterotrophic feeding.

Anthocaulus A specialized tissue that forms at the base of a polyp, aiding in the regeneration of new polyps and potentially contributing to colony propagation.

Aragonite Saturation State (Ω_arag) A measure of the concentration of carbonate ions in seawater relative to aragonite, a form of calcium carbonate. It is critical for understanding the conditions favorable for coral calcification.

Asexual Reproduction Reproductive process resulting in clonal offspring, including budding, fragmentation, and polyp bail-out, crucial for the growth and propagation of coral colonies.

Atmospheric Correction A preprocessing step that removes atmospheric effects (e.g., haze, scattering, absorption) from satellite imagery to retrieve true surface reflectance values.

AutoCAD® Map A GIS-enabled version of AutoCAD software used for vector-based digitization of features such as coral patches from orthophoto mosaics.

Axial Corallite The central corallite in a branching coral from which new branches emerge. It is usually larger and more prominent than radial corallites.

Benthic Habitat Mapping The identification and classification of seafloor habitats, such as corals, sand, rubble, and seaweeds, using remote sensing and field data.

Bioerosion The process through which organisms such as boring sponges, algae, and other invertebrates break down coral skeletons, contributing to the dynamic nature of reef ecosystems.

Calcification The biological process by which corals deposit calcium carbonate to form their skeletons, influenced by various environmental factors including pH, temperature, and the availability of carbonate ions.

Calice (Calyx) The skeletal cup or depression in which a single coral polyp sits. Each calice is created by the polyp secreting calcium carbonate.

Coenosarc The living tissue that connects individual polyps in a colony, facilitating nutrient and symbiont sharing among polyps.

Coenosteum The common skeletal structure that connects individual corallites within a colonial coral. It can be solid or porous, and its texture and appearance can be distinctive for different coral taxa.

Columella The central, often rod-like skeletal structure within a corallite that supports the polyp's tissue. The structure and presence of a columella can be crucial for species identification.

Confusion Matrix A table used in accuracy assessments of classification models, comparing predicted and observed class labels.

Coral Bleaching A stress response in which corals expel their zooxanthellae, leading to a loss of color and, if prolonged, potentially resulting in coral mortality due to the loss of the symbiotic relationship.

Corallite The individual skeletal cup produced by a single coral polyp. Corallites may be solitary or part of a colonial structure and vary widely in size, shape, and structure.

Corallum The entire skeletal structure of a coral colony, composed of multiple corallites and the coenosteum.

Costae The external skeletal ridges that correspond to the internal septa. They extend beyond the corallite wall and can help in distinguishing between species.

Costae The ridges or lines on the external surface of the corallite wall, corresponding to the septa within the calice and often used in species identification.

Digital Number (DN) The raw pixel value recorded by a satellite sensor, which must be converted to radiance or reflectance for analysis.

M. R. Shokri, K. Kabiri, *Coral Reefs of Iran*, Coral Reefs of the World 22, https://doi.org/10.1007/978-3-032-00598-4

Drone-Based Mapping The use of unmanned aerial vehicles (UAVs) equipped with cameras to acquire high-resolution imagery for environmental mapping, especially effective in shallow coastal waters.

Edwards & Mumby Method A technique used for correcting water column effects in satellite imagery by estimating attenuation coefficients for spectral bands.

ENVI™ A software suite used for processing and analyzing geospatial imagery, particularly for remote sensing applications.

Epitheca A thin, outer skeletal layer secreted by corals, often covering the coenosteum and providing protection to the underlying tissue.

FLAASH (Fast Line-of-sight Atmospheric Analysis of Spectral Hypercubes) An atmospheric correction tool in ENVI that converts radiance to surface reflectance using radiative transfer models.

Foliaceous Describes a growth form that resembles leaves or flattened plates, often found in corals like Montipora and Echinopora.

GeoEye-1/QuickBird/WorldView-2 High-resolution commercial satellites commonly used in coral reef mapping due to their fine spatial resolution and multispectral capabilities.

Geometric Correction The process of aligning satellite or aerial images with geographic coordinates using ground control points to ensure spatial accuracy.

Gonochorism Gonochorism is a sexual system where there are two sexes and each individual organism is either male or female.

Ground Control Points (GCPs) Reference points with known geographic coordinates used to georeference aerial or satellite images for spatial accuracy.

Hermaphroditic Refers to coral species possessing both male and female reproductive organs, capable of producing both eggs and sperm.

Hermatypic Corals Reef-building corals that contain zooxanthellae, primarily found in shallow, sunlit waters where conditions are optimal for photosynthesis.

Holobiont The ecological unit comprising the coral host and its associated microbiota, including bacteria, archaea, viruses, fungi, and the zooxanthellae, all of which play integral roles in coral health and function.

Intramural Budding A type of asexual reproduction where new polyps form within the wall of an existing polyp, leading to the expansion of the colony.

Intratentacular Budding A form of asexual reproduction where new polyps develop within the parental corallite, leading to the formation of new corallites.

Lyzenga Method A semi-analytical technique to derive relative bathymetry from multispectral imagery based on substrate reflectance and water depth.

Mesenterial Filaments Thread-like structures within the coral polyp's gut cavity used for digestion, containing nematocysts and enzymes to capture and break down prey.

Mesenteries Internal radial partitions within the polyp's body that support the tentacles and digestive organs. The number and arrangement of mesenteries can be diagnostic.

Nematocysts Specialized stinging cells located on the tentacles and surface of the polyp, used for capturing prey and defense.

Orthophoto Mosaic A geometrically corrected image composed of multiple overlapping photos stitched together, often used for mapping benthic habitats.

Paliform Lobes Secondary skeletal structures that project inward from the septa towards the columella, aiding in polyp support and structural complexity.

Pix4Dcapture® A mobile application used to program and control drone flights for automated, high-resolution data collection.

Planimetric Accuracy The accuracy of the horizontal positioning of features in a map or image.

Planula The free-swimming larval stage of corals, resulting from sexual reproduction, which eventually settles and metamorphoses into a new polyp.

Polyp Bail-Out A stress-induced escape mechanism where individual polyps detach from the skeleton and free-float to settle in a new location, potentially forming new colonies.

Radial Corallites Smaller corallites that protrude from the sides of the branch or colony, often used in taxonomic differentiation.

Random Forest Classification A machine learning algorithm used for supervised image classification that operates by constructing multiple decision trees.

Remote Sensing The acquisition of information about Earth's surface features without direct contact, typically via satellite or airborne sensors.

Seafloor Reflectance The amount of light reflected from the ocean bottom, as recorded by remote sensing platforms; important in habitat classification.

Septa Vertical skeletal elements within the corallite that radiate from the columella towards the corallite wall. Their arrangement, thickness, and ornamentation are important for identification.

Septocostae Internal and external skeletal elements that form continuous ridges from the corallite wall, combining features of septa and costae.

Shallow-Water Mapping Mapping of underwater features in shallow coastal zones (typically <10 m depth), where light penetration allows for optical methods.

Spectral Bands Discrete wavelength intervals in which satellite sensors capture data; used to differentiate between surface features based on their reflectance properties.

Supervised Classification An image classification method where the user provides training data to guide the algorithm in assigning land cover classes.

Synapticulae Horizontal skeletal elements that connect adjacent septa within the corallite, adding structural integrity and complexity.

Theca The outer wall of a corallite, providing support and protection to the polyp. The thickness, texture, and presence of the theca are key identification features.

Topological Map A spatial dataset where geographic features are digitized and categorized into defined classes (e.g., coral types), maintaining relationships among them.

Trabeculae Minute, rod-like skeletal structures that form the basic building blocks of the coral skeleton, contributing to its porous and lightweight nature.

UAV (Unmanned Aerial Vehicle) An aircraft piloted without a human onboard, commonly referred to as a drone, used for high-resolution image acquisition in marine and coastal studies.

UTM (Universal Transverse Mercator) A geographic coordinate system used to map data in meters, providing uniform spatial referencing.

Water Column Correction A technique used to reduce the effects of water depth and turbidity on satellite images to improve seafloor mapping accuracy.

Zonation The spatial distribution of different coral species or growth forms across different reef zones, influenced by factors such as light, wave action, and water depth. Understanding zonation can aid in the identification of species and their ecological roles.

Zooxanthellae Symbiotic dinoflagellates of the genus *Symbiodinium* that live within coral tissues, providing essential nutrients through photosynthesis and contributing to coral calcification.

References

Ahmed H, Mohammed H, Ahmed A, Al-Waeli A, Abbas M, Ali M (2023) First record of the stony coral paracyathus stokesii Edward and Haime, 1848 (Caryophylliidae) in the Eastern Breakwater of the New Al-Faw Grand Port, Basrah, Iraq

Aminrad T, Azini MR (2013) New and previous records of scleractinian corals from Chabahar Bay, Sistan & Baluchistan, Iran. Ecopersia 1(4):407–418

Baird A, Morse A (2004) Induction of metamorphosis in larvae of the brooding corals Acropora palifera and Stylophora pistillata. Mar Freshw Res. https://doi.org/10.1071/MF03121

Bargahi HR, Shokri MR, Kaymaram F, Fatemi MR (2020) Changes in reef fish assemblages following multiple bleaching events in the world's warmest sea (Kish Island, the Persian Gulf). Coral Reefs 39:603–624. https://doi.org/10.1007/s00338-020-01945-3

Barnes RSK, Hughes RN (1999) An introduction to marine ecology, 3rd edn. Oxford Press, Oxford

Basson P, Burchard JH, Hardy JT, Price A (1977) Biotopes of the prevention, western Arabian Gulf. Aramco Department of Loss and Environmental Affairs, Dhahran

Bauman AG, Pratchett MS, Baird AH, Riegl B, Heron SF, Feary DA (2013) Variation in the size structure of corals is related to environmental extremes in the Persian Gulf. Mar Environ Res 84:43–50. https://doi.org/10.1016/j.marenvres.2012.11.007

Benzoni F (2006) Psammocora albopicta sp nov., a new species of Scleractinian Coral from the Indo-West Pacific (Scleractinia; Siderastreidae). Zootaxa 1358:49–57. https://doi.org/10.11646/zootaxa.1358.1.3

Benzoni F, Stefani F, Stolarski J, Pichon M, Mitta G, Galli P (2007) Debating phylogenetic relationships of the scleractinian Psammocora: Molecular and morphological evidences. Contrib to Zool Bijdr tot Dierkd 76:35–54. https://doi.org/10.1163/18759866-07601004

Benzoni F, Stefani F, Pichon M, Galli P (2010) The name game: Morpho-molecular species boundaries in the genus Psammocora (Cnidaria, Scleractinia). Zool J Linn Soc 160:421–456. https://doi.org/10.1111/j.1096-3642.2010.00622.x

Benzoni F, Arrigoni R, Stefani F, Pichon M (2011) Phylogeny of the coral genus Plesiastrea (Cnidaria, Scleractinia). Contrib to Zool Bijdr tot Dierkd 80:231–249. https://doi.org/10.1163/18759866-08004002

Brewer PG, Dyrssen D (1985) Chemical oceanography of the Persian Gulf. Prog Oceanogr 14:41–55. https://doi.org/10.1016/0079-6611(85)90004-7

Buchanan JR, Krupp F, Burt JA, Feary DA, Ralph GM, Carpenter KE (2016) Living on the edge: vulnerability of coral-dependent fishes in the Gulf. Mar Pollut Bull 105:480–488. https://doi.org/10.1016/j.marpolbul.2015.11.033

Budd ANNF, Fukami H, Smith ND, Knowlton N (2012) Taxonomic classification of the reef coral family Mussidae (Cnidaria: Anthozoa: Scleractinia). Zool J Linn Soc 166:465–529. https://doi.org/10.1111/j.1096-3642.2012.00855.x

Burchard JE (1979) Coral fauna of the western Persian Gulf. ARAMCO, Dhahran

Burt J, Bartholomew A, Usseglio P (2008) Recovery of corals a decade after a bleaching event in Dubai, United Arab Emirates. Mar Biol 154:27–36. https://doi.org/10.1007/s00227-007-0892-9

Carpenter KE, Harrison PL, Hodgson G, Alsaffar AH, Alhazeem SH (1997) The corals and coral reef fishes of Kuwait. Biological sciences faculty books, 7. https://digitalcommons.odu.edu/biology_books/7

Chevalier JL (1971) Les scleractiniaires de la Melanesie Francaise (Nouvelle Caledonie, Iles Chesterfi eld, Iles Loyaute, Nouvelles Hebrides). 1ere Partie. In: Exped Francaise recifs coralliens Nouvelle Caledonie, vol 5. Fond Singer Polignac. Paris, pp 5–307

Chiappone M, Sealey K, Lott C (1996) Hermatypic scleractinian corals of the Southeastern Bahamas: a comparison to western Atlantic reef systems. Caribb J Sci 32:1–13

Claereboudt MR (2006) Reef Corals and Coral Reefs of the Gulf of Oman. The Historical Association of Oman

Coles SL (1988) Limitations of reef coral development in the Persian Gulf: temperature or algal competition? In: Proceedings of the sixth international coral reef symposium, vol 3, pp 211–216

Coles S (1996) Corals of Oman, p 125

Coles SL (2003) Coral species diversity and environmental factors in the Arabian Gulf and the Gulf of Oman: a comparison to the Indo-Pacific region. Atoll Res Bull:1–19. https://doi.org/10.5479/si.00775630.507.1

Coles SL, Fadlallah YH (1991) Reef coral survival and mortality at low temperatures in the Arabian Gulf: new species-specific lower temperature limits. Coral Reefs 9:231–237. https://doi.org/10.1007/BF00290427

Costanza R, d'Arge R, de Groot R, Farber S, Grasso M, Hannon B, Limburg K, Naeem S, O'Neill RV, Paruelo J, Raskin RG, Sutton P, van den Belt M (1997) The value of the world's ecosystem services and natural capital. Nature 387:253–260. https://doi.org/10.1038/387253a0

Crabbe MJC, Smith DJ (2005) Sediment impacts on growth rates of Acropora and Porites corals from fringing reefs of Sulawesi, Indonesia. Coral Reefs 24:437–441. https://doi.org/10.1007/s00338-005-0004-6

Darkhal H, Kabiri K, Shokri MR (2025) Optimizing coral reef restoration: A GIS-based site selection for coral nurseries in Kish Island, the Persian Gulf. J Nat Conserv 86:126898. https://doi.org/10.1016/j.jnc.2025.126898

DeVantier L, Turak E (2017) Species richness and relative abundance of reef-building corals in the Indo-West Pacific. Diversity 9

Downing N (1995) Coral reef communities in an extreme environment: the Northwest Arabian Gulf. In: Proceedings of the fifth international coral reef symposium, vol 6, pp 343–348

Elliott AJ, Savidge G (1990) Some features of the upwelling off Oman. J Mar Res 48:319–333

Fadlallah YH, Allen KW, Estudillo RA (1995) Mortality of shallow reef corals in the western Arabian Gulf following aerial exposure in winter. Coral Reefs 14:99–107. https://doi.org/10.1007/BF00303430

M. R. Shokri, K. Kabiri, *Coral Reefs of Iran*, Coral Reefs of the World 22, https://doi.org/10.1007/978-3-032-00598-4

Fatemi SMR, Shokri MR (2001) Iranian coral reefs status with particular reference to Kish Island, Persian Gulf. In: Indian Ocean regional workshop, p 13

Fawzi NA-M, Fieseler CM, Helmuth B, Leitão A, Al-Ainsi M, Al Mukaimi M, Al-Saidi M, Al Senafi F, Bejarano I, Ben-Hamadou R, D'Addario J, Mohamed AMD, Giraldes BW, Glowka L, Johnson MD, Lyons BP, Mateos-Molina D, Marshall CD, Mohammed S, Range P, Shokri MR, Wong JMK, Pyenson ND (2022) Diplomacy for the world's hottest sea. Science 376:1389–1390. https://doi.org/10.1126/science.add1555

Fieseler CM, Al-Mudaffar Fawzi N, Helmuth B, Leitão A, Al Ainsi M, Al Mukaimi M, Al-Saidi M, Al Senafi F, Bejarano I, Ben-Hamadou R, D'Addario J, Mujthaba Dheen Mohamed A, Giraldes BW, Glowka L, Johnson MD, Lyons BP, Mateos-Molina D, Marshall CD, Mohammed S, Range P, Reza Shokri M, Wong JMK, Pyenson ND (2023) Expanding ocean protection and peace: a window for science diplomacy in the Gulf. R Soc Open Sci 10:230392. https://doi.org/10.1098/rsos.230392

Fukami H, Chen CA, Budd AF, Collins A, Wallace C, Chuang Y-Y, Chen C, Dai C-F, Iwao K, Sheppard C, Knowlton N (2008) Mitochondrial and nuclear genes suggest that stony corals are monophyletic but most families of stony corals are not (Order Scleractinia, Class Anthozoa, Phylum Cnidaria). PLoS One 3:e3222

Ghazilou A (2019) Bleaching of corals in Chah Bahar Bay, Sea of Oman. Tehran

Goffredo S, Dubinsky ZVY (2016) The Cnidaria, past, present and future: the world of Medusa and her sisters. Springer International Publishing

Grandcourt E (2012) In: Riegl BM, Purkis SJ (eds) Reef fish and fisheries in the Gulf BT – Coral Reefs of the Gulf: adaptation to climatic extremes. Springer, Dordrecht, pp 127–161

Harger JRE (1984) Rapid survey techniques to determine distribution and structure of coral communities. In: Comparing coral reef survey methods. UNEPUNESCO Workshop, Thailand. Phuke, Thiland, pp 83–91

Harrison PL, Babcock RC, Bull GD, Oliver JK, Wallace CC, Willis BL (1984) Mass spawning in tropical reef corals. Science 223:1186–1189. https://doi.org/10.1126/science.223.4641.1186

Heyward AJ, Babcock RC (1986) Self- and cross-fertilization in scleractinian corals. Mar Biol 90:191–195. https://doi.org/10.1007/BF00569127

Hodgson G, Carpenter K (1995) Scleractinian corals of Kuwait. Pacific Sci 49:227–246

Hoegh-Guldberg O, Andréfouët S, Fabricius K, Diaz-Pulido G, Lough J, Marshall P, Pratchett MS (2011) Vulnerability of coral reefs in the tropical Pacific to climate change, pp 251–296

Hoeksema BW, Cairns S (2024a) world List of Scleractinia. Fungiidae Dana, 1846. Accessed through: World Register of Marine Species at: https://www.marinespecies.org/aphia.php?p=taxdetails&id=196100

Hoeksema BW, Cairns S (2024b) World list of Scleractinia. Oxypora Saville Kent, 1871. Accessed through: World Register of Marine Species at: https://www.marinespecies.org/aphia.php?p=taxdetails&id=205536

Kabiri K, Pradhan B, Samimi-Namin K, Moradi M (2013) Detecting coral bleaching, using QuickBird multi-temporal data: a feasibility study at Kish Island, the Persian Gulf. Estuar Coast Shelf Sci 117:273–281. https://doi.org/10.1016/j.ecss.2012.12.006

Kabiri K, Rezai H, Moradi M (2018) Mapping of the corals around Hendorabi Island (Persian Gulf), using WorldView-2 standard imagery coupled with field observations. Mar Pollut Bull 129:266–274. https://doi.org/10.1016/j.marpolbul.2018.02.045

Kabiri K, Rezai H, Moradi M (2020) A drone-based method for mapping the coral reefs in the shallow coastal waters – case study: Kish Island, Persian Gulf. Earth Sci Informatics. https://doi.org/10.1007/s12145-020-00507-z

Kassler P (1973) The structural and geomorphic evolution of the Persian Gulf. In: Purser BH (ed) The Persian Gulf: Holocene carbonate sedimentation and diagenesis in a shallow epicontinental sea. Springer, Berlin, Heidelberg, pp 11–32

Kavousi J, Seyfabadi J, Rezai H, Fenner D (2011) Coral reefs and communities of Qeshm Island, the Persian Gulf. Zool Stud 50:276–283

Kitahara MV, Cairns SD, Stolarski J, Blair D, Miller DJ (2010) A comprehensive phylogenetic analysis of the scleractinia (Cnidaria, Anthozoa) based on mitochondrial CO1 sequence data. PLoS One 5:e11490

Kitano YF, Benzoni F, Arrigoni R, Shirayama Y, Wallace CC, Fukami H (2014) A phylogeny of the Family Poritidae (Cnidaria, Scleractinia) based on molecular and morphological analyses. PLoS One 9:e98406

Lincoln S, Buckley P, Howes EL, Maltby KM, Pinnegar JK, Ali TS, Alosairi Y, Al-ragum A, Baglee A, Balmes CO, Ben HR, Burt JA, Claereboudt M, Glavan J, Mamiit RJ, Abdulla H, Naser A, Sedighi O, Shokri MR, Shuhaibar B (2021) A regional review of marine and coastal impacts of climate change on the ROPME sea area

Lyzenga D (1978) Passive remote sensing techniques for mapping water depth and bottom features. Appl Opt 17:379–383. https://doi.org/10.1364/AO.17.000379

Maghsoudlou A (2008) Hard Corals of the Iranian coastal waters of the Persian Gulf. Iranian National Institute for Oceanography and Atmospheric Science (INIOAS)

Maghsoudlou A (2020) Check list of Iran marine Cnidarians (Animalia, Cnidaria). Iran J Anim Biosyst 16:1–12. https://doi.org/10.22067/ijab.v16i1.85363

Maltby KM, Howes EL, Lincoln S, Pinnegar JK, Buckley P, Ali TS, Al Balushi B, Al Ragum A, Al Shukail HSA, Balmes CO, Ben-Hamadou R, Claereboudt MRG, Mamiit RJE, Naser HA, Shokri MR, Le Quesne WJF (2022) Marine climate change risks to biodiversity and society in the ROPME Sea Area. Clim Risk Manag 35:100411. https://doi.org/10.1016/j.crm.2022.100411

Miller KJ (1992) Morphological variation in the scleractinian coral Platygyra daedalea (Ellis and Solander, 1786), genetically or environmentally determined? In: Proceedings of the seventh international coral reef symposium, vol 1, pp 550–556

Miller K, Babcock R (1997) Conflicting morphological and reproductive species boundaries in the coral genus platygyra. Biol Bull 192:98–110. https://doi.org/10.2307/1542579

Miller K, Benzie J (1997) no clear genetic distinction between morphological species within the coral genus Platygyra. Bull Mar Sci 61:907–917

Moghaddam S, Shokri MR, Tohidfar M (2021) The enhanced expression of heat stress-related genes in scleractinian coral 'Porites harrisoni' during warm episodes as an intrinsic mechanism for adaptation in 'the Persian Gulf'. Coral Reefs 40:1013–1028. https://doi.org/10.1007/s00338-021-02100-2

Moradi M, Malekzadeh-Viayeh R, Eshaghi-Rad J (2014) Biodiversity of scleractinian corals in the reefs of Qeshm and Larak Islands of the Persian Gulf, in association with environmental variables. J Mar Biol Assoc United Kingdom 94:907–916. https://doi.org/10.1017/S0025315414000411

Morrison J (1997) Inter-monsoonal changes in the T-S properties of the near-surface waters of the Northern Arabian Sea. Geophys Res Lett Geophys RES LETT 24:2553–2556. https://doi.org/10.1029/97GL01876

Mumby PJ, Edwards AJ (2000) Water column correction approaches. In: Green EP, Mumby PJ, Edwards AJ, Clark CD (eds) Remote sensing handbook for tropical coastal management. Unesco, Paris, p 316

Orrell T (2019) NMNH extant specimen records. Version 1.21. National Museum of Natural History, Smithsonian Institution. Occurrence dataset.

Pirazzoli PA, Reyss J-L, Fontugne M, Haghipour A, Hilgers A, Kasper HU, Nazari H, Preusser F, Radtke U (2004) Quaternary coral-reef terraces from Kish and Qeshm Islands, Persian Gulf: new radiometric ages and tectonic implications. Quat Int 120:15–27. https://doi.org/10.1016/j.quaint.2004.01.003

Pratchett M, Baird A, Bauman A, Burt J (2016) Abundance and composition of juvenile corals reveals divergent trajectories for coral assemblages across the United Arab Emirates. Mar Pollut Bull. https://doi.org/10.1016/j.marpolbul.2016.11.036

Purkis S, Riegl B (2005) Spatial and temporal dynamics of Arabian Gulf coral assemblages quantified from remote-sensing and in situ monitoring data. Mar Ecol Ser 287:99–113. https://doi.org/10.3354/meps287099

Purkis SJ, Riegl BM (2012) Geomorphology and reef building in the SE Gulf. In: Riegl BM, Purkis SJ (eds) BT coral reefs of the Gulf: adaptation to climatic extremes. Springer, Dordrecht, pp 33–50

Rahmani M, Rahimian H (2013) Preliminary study on Acropora (Scleractinia: Astrocoeniina: Acroporidae) of the Persian Gulf, with emphasis on the north and northeastern areas. Turkish J Zool 37:308–320. https://doi.org/10.3906/zoo-1112-21

Reynolds RM (1993) Physical oceanography of the Persian Gulf, Strait of Hormuz, and the Gulf of Oman – results from the Mt Mitchell expedition. Mar Pollut Bull 27:35–59

Rezai H, Wilson S, Claereboudt M, Riegl B (2004) Coral reef status in the ROPME sea area: Arabian/Persian Gulf, Gulf of Oman and Arabian Sea. Status of Coral Reefs of the World, pp 155–170.

Riegl B (1999) Corals in a non-reef setting in the southern Arabian Gulf (Dubai, UAE): fauna and community structure in response to recurring mass mortality. Coral Reefs 18:63–73. https://doi.org/10.1007/s003380050156

Riegl B (2001) Inhibition of reef framework by frequent disturbance: examples from the Arabian Gulf, South Africa, and the Cayman Islands. Palaeogeogr Palaeoclimatol Palaeoecol 175:79–101. https://doi.org/10.1016/S0031-0182(01)00387-X

Riegl B (2003) Climate change and coral reefs: different effects in two high-latitude areas (Arabian Gulf, South Africa). Coral Reefs 22:433–446. https://doi.org/10.1007/s00338-003-0335-0

Riegl BM, Purkis SJ (2012a) Coral reefs of the Gulf: adaptation to climatic extremes in the World's Hottest Sea. In: Riegl BM, Purkis SJ (eds) BT – coral reefs of the Gulf: adaptation to climatic extremes. Springer, Dordrecht, pp 1–4

Riegl BM, Purkis SJ (2012b) Environmental constraints for reef building in the Gulf. In: Riegl BM, Purkis SJ (eds) BT – coral reefs of the Gulf: adaptation to climatic extremes. Springer, Dordrecht, pp 5–32

Riegl BM, Purkis SJ, Al-Cibahy AS, Abdel-Moati MA, Hoegh-Guldberg O (2011) Present limits to heat-adaptability in corals and population-level responses to climate extremes. PLoS One. https://doi.org/10.1371/journal.pone.0024802

Riegl BM, Benzoni F, Samimi-Namin K, Sheppard C (2012) The hermatypic scleractinian (hard) coral fauna of the Gulf. In: Coral reefs of the Gulf: adaptation to climatic extremes. Springer, Dordrecht, pp 187–224

Riegl B, Johnston M, Purkis S, Howells E, Burt J, Steiner SCC, Sheppard CRC, Bauman A (2018) Population collapse dynamics in Acropora downingi, an Arabian/Persian Gulf ecosystem-engineering coral, linked to rising temperature. Glob Chang Biol 24:2447–2462. https://doi.org/10.1111/gcb.14114

Rohwer F, Seguritan V, Azam F, Knowlton N (2002) Diversity and distribution of coral-associated bacteria. Marine Ecol Prog Ser 243:1–10

Ross DA, Uchupi E, White RS (1986) The geology of the Persian Gulf-Gulf of Oman Region: a synthesis. Rev Geophys 24:537–556. https://doi.org/10.1029/RG024i003p00537

Sadeghi P, Loghmani M (2009) New observation of three species of hard coral from Chabahar Bay (Oman Sea), Iran. J Biol Sci 9:837–840. https://doi.org/10.3923/jbs.2009.837.840

Sale PF, Feary DA, Burt JA, Bauman AG, Cavalcante GH, Drouillard KG, Kjerfve B, Marquis E, Trick CG, Usseglio P, Van Lavieren H (2011) The growing need for sustainable ecological management of marine communities of the Persian Gulf. Ambio 40:4–17. https://doi.org/10.1007/s13280-010-0092-6

Salimi MA, Mostafavi PG, Fatemi SMR, Aeby GS (2017) Health status of corals surrounding Kish Island, Persian Gulf. Dis Aquat Organ 124:77–84. https://doi.org/10.3354/dao03105

Salimi PA, Chen PGMCA, Fatemi SMR, Pichon M (2018) The Scleractinia (Cnidaria: Anthozoa) of Abu-Musa and Sirri Islands, Persian Gulf

Samimi-Namin K, Ofwegen L (2009) Some shallow water octocorals (Coelenterata: Anthozoa) of the Persian Gulf. Zootaxa:1–52. https://doi.org/10.5281/zenodo.186743

Schils T, Wilson SC (2006) Temperature threshold as a biogeographic barrier in northern Indian Ocean macroalgae. J Phycol 42:749–756. https://doi.org/10.1111/j.1529-8817.2006.00242.x

Seibold E, Diester L, Fütterer D, Lange H, Müller P, Werner F (1973) Holocene sediments and sedimentary processes in the Iranian part of the Persian Gulf. In: Purser BH (ed) The Persian Gulf: Holocene carbonate sedimentation and diagenesis in a shallow epicontinental sea. Springer, Heidelberg, pp 57–80

Sharif Ranjabr M, Askari Hesni M, Ghavasi M, Dab K, Chegonian A (2015) Integrated coral reef management program of Tonb, Abu Musa, and Siri Islands by hard corals species identification, health status control and distribution mapping. Tehran

Sheppard CRC (1988) Similar trend, different causes: responses of corals to stressed environments in Arabian seas. In: Proceedings of the 6th international coral reef congress, Townsville, vol 3, pp 297–302

Sheppard C (2016) Coral reefs in the Gulf are mostly dead now, but can we do anything about it? Mar Pollut Bull 105:593–598. https://doi.org/10.1016/j.marpolbul.2015.09.031

Sheppard C, Salm R (1988) Reef and coral communities of Oman, with a description of a new coral species (Order Scleractinia, genus Acanthastrea). J Nat Hist 22:263–279. https://doi.org/10.1080/00222938800770201

Sheppard CR, Sheppard ALS (1991) Corals and coral communities of Arabia. In: Buttiker W, Krupp F (eds) Fauna of Saudi Arabia, vol 12. Natural History Museum, Basle, pp 3–170

Sheppard CRC, Sheppard ALS (1993) Corals and coral communities of Arabia. In: Fauna of Saudi Arabia, pp 12, 170

Sheppard C, Price A, Roberts C (1992) Marine ecology of the Arabian Region: patterns and processes in extreme tropical environments, 1st edn. Academic Press

Sheppard C, Al-Husiani M, Al-Jamali F, Al-Yamani F, Baldwin R, Bishop J, Benzoni F, Dutrieux E, Dulvy NK, Durvasula SRV, Jones DA, Loughland R, Medio D, Nithyanandan M, Pilling GM, Polikarpov I, Price ARG, Purkis S, Riegl B, Saburova M, Namin KS, Taylor O, Wilson S, Zainal K (2010a) The Gulf: a young sea in decline. Mar Pollut Bull 60:13–38. https://doi.org/10.1016/j.marpolbul.2009.10.017

Sheppard C, Al-Husiani M, Al-Jamali F, Al-Yamani F, Baldwin R, Bishop J, Benzoni F, Dutrieux E, Dulvy N, Durvasula S, Jones D, Loughland R, Medio D, Manickam N, Pilling G, Polikarpov I, Price A, Purkis S, Riegl B, Zainal K (2010b) The Persian/Arabian Gulf: a young sea in decline. Mar Pollut Bull 60:13–38. https://doi.org/10.1016/j.marpolbul.2009.10.017

Shinn EA (1976) Coral reef recovery in Florida and the Persian Gulf. Environ Geol 1:241–254. https://doi.org/10.1007/BF02407510

Shojae F, Kamrani E, Shokri MR, Ranjbar MS, Moradi M, Hesni MA (2010) New records of three hard coral species from north-east of Larak Island (Persian Gulf, Iran). Mar Biodivers Rec 3:1–6. https://doi.org/10.1017/s1755267210000515

Shokri MR (2001) Status of coral reefs around farur island, iran (persian gulf) final report 2001, pp 1–35

Shokri MR (2020) Status and trends of coral reefs in the ROPME sea area: past, present and future. Kuwait

Shokri MR, Mohammadi M (2021) Effects of recreational SCUBA diving on coral reefs with an emphasis on tourism suitability index and carrying capacity of reefs in Kish Island, the northern Persian Gulf. Reg Stud Mar Sci 45:101813. https://doi.org/10.1016/j.rsma.2021.101813

Shokri MR, Oladi M (2021) New record of azooxanthellate dendrophyllia cribrosa in the Persian gulf (Anthozoa, scleractinia, dendrophylliidae). Spixiana 44:11–16

Shokri MR, Ardakani O, Sharifi A, Abdoullahi P, Nazarian M (2000) Status of coral reefs around Kish Island in the Iranian Persian Gulf. In: Proceedings of the international symposium on the extent and impact of coral bleaching in the Arabian Region, pp 5–9

Spalding MD, Ravilious C, Green EP (2001) World atlas of coral reefs, 1st edn. UNEP World Conservation Monitoring Centre. University of California Press, Berkley, USA

Supreme Council for Environment and Natural Reserves of Qatar (SCENR) (2007) Sensitivity mapping of the eastern coast of Qatar – phase, vol 1. Doha

Tavakoli-Kolour P, Kavousi J, Rezai H (2015) Outbreak of growth anomalies in coral communities of Qeshm Island, Persian Gulf. Int Aquat Res 7:151–156. https://doi.org/10.1007/s40071-015-0100-3

Uchupi E, Swift SA, Ross DA (1996) Gas venting and late Quaternary sedimentation in the Persian (Arabian) Gulf. Mar Geol 129:237–269. https://doi.org/10.1016/0025-3227(96)83347-0

Vajed Samiei J, Dab K, Ghezellou P, Shirvani A (2013) Some scleractinian corals (Scleractinia: Anthozoa) of Larak Island, Persian Gulf. Zootaxa 3636:101–143. https://doi.org/10.11646/zootaxa.3636.1.5

Vaughan TW, Wells JW (1943) Revision of the suborders families, and Genera of the Scleractinia. Geological Society of America

Venkataraman K, Satyanarayana C (2012) Coral Identification manual. Zoological Survey of India, Kolkata

Veron JEN (1986) Corals of Australia and the Indo-Pacific. Angus and Robertson, Sydney

Veron JEN (2000) Corals of the world. Australian Institute of Marine Science, Townsville

Veron JEN, Pichon M (1976) Scleractinia of Eastern Australia. Australian Govt. Pub. Service, Canberra

Veron JEN, Pichon M, Wijsman-Best M (1977) Scleractinia of Eastern Australia, II: families Faviidae, Trachyphylliidae. AIMS Monogr Ser 3:227

Veron JEN, John EN, Wallace CC (1984) Scleractinia of eastern Australia. Part V. Family Acroporidae. Scleractinia East Aust Part V 6:1–485

Vogt H (1996) Investigations on coral reefs in the Jubail wildlife sanctuary using under water video recordings and digital image analysis. In: Krupp F, Abuzinada AH, Nader IA (eds) A marine wildlife sanctuary for the Persian Gulf. NCWCD Riyadh and Senckenbergische Naturforschende Gesellschaft, Frankfurt, pp 302–327

Wallace CC (1978) The coral genus Acropora (Scleractinia: As_trocoeniina; Acroporidae) in the central and southern Great Barrier Reef. Mem Queensl Museum 18:273–319

Wallace C (1999) Staghorn corals of the world. CSIRO Publishing

Weil E, Hammerman NM, Becicka RL, Cruz-Motta JJ (2020) Growth dynamics in Acropora cervicornis and A. prolifera in southwest Puerto Rico. PeerJ 2020:1–23. https://doi.org/10.7717/peerj.8435

Wells JW (1956) Scleractinia. In: Moore R (ed) Treatise on invertebrate paleontology. Part F. Coelenterata. University Kansas Press, Lawrence, pp 328–444

Wilson S, Fatemi SMR, Shokri MR, Claereboudt M (2002) Status of coral reefs of the Persian/Arabian Gulf and Arabian Sea Region. Status of coral reefs of the world, pp 53–62

WoRMS Editorial Board (2004) World register of marine species. In: VLIZ. https://www.marinespecies.org

Yap HT, Gomez ED (1984) Growth of Acropora pulchra. Mar Biol 81:209–215. https://doi.org/10.1007/BF00393119

General Index

M. R. Shokri, K. Kabiri, *Coral Reefs of Iran*, Coral Reefs of the World 22, https://doi.org/10.1007/978-3-032-00598-4

Species Index

M. R. Shokri, K. Kabiri, *Coral Reefs of Iran*, Coral Reefs of the World 22, https://doi.org/10.1007/978-3-032-00598-4

MIX
Papier aus verantwortungsvollen Quellen
Paper from responsible sources
FSC® C105338

If you have any concerns about our products,
you can contact us on
ProductSafety@springernature.com

In case Publisher is established outside the EU,
the EU authorized representative is:
Springer Nature Customer Service Center GmbH
Europaplatz 3, 69115 Heidelberg, Germany

Printed by Libri Plureos GmbH
in Hamburg, Germany